Property of
The East Everglades Orchid Society (5-03)
Homestead FL

The Scent of Orchids

Roman Kaiser

The Scent of Orchids

Olfactory and chemical investigations

ELSEVIER

Amsterdam – London – New York – Tokyo

Editiones Roche, Basel

1993

Frontispiece: *Cattleya labiata*
See Figure 55

ELSEVIER SCIENCE PUBLISHERS B.V.
Molenwerf 1
P.O. Box 211, 1000 AE Amsterdam, The Netherlands

Address of the author:
Roman Kaiser
Givaudan-Roure Research Ltd.
CH-8600 Dübendorf

Photographs by the author (exceptions noted).

Layout: Martin Schneider

English translation: BMP Translation Services, Basel
Lithographer: Photolitho Sturm AG, Muttenz
Composed and printed by Morf & Co. AG, Basel
Proofreading by Rotstift AG, Basel
Binding by Buchbinderei Grollimund AG, Reinach
Printed on Biberist GS, double-coated, chlorine-free, super-white, wood-free, mat

German edition: Editiones Roche, ISBN 3-907946-87-1

71314

ISBN 0-444-89841-7

© 1993 Editiones Roche
F. Hoffmann-La Roche AG, Basel

No part of this publication may be reproduced or transmitted in any form or by any means, electronic, mechanical, photocopying, recording or otherwise, without the prior written of Editiones Roche, F. Hoffmann-La Roche AG, CH-4002 Basel

Contents

Introduction 9

Part One: Introduction to the World of Orchid Scents 11

A few comments on the nature of orchids 13

Plant scents 17

Trapping and investigation of orchid scents 22

Scent and pollination principles 29

Verbal description of scents 40

Part Two: Interdisciplinary Discussion of Orchid Scents 47

Orchids of the American tropics 49

Orchids of the African tropics 123

Orchids of the Indo-Australian tropics and subtropics 144

Some European orchids 176

Part Three: The Chemistry of Orchid Scents 183

General remarks 185

Analytical composition of the individual orchid scents 188

Appendix 247

References 249

Suggestions for further reading 253

Index of orchid species discussed 255

Acknowledgements 258

Foreword

I must admit that when Roman Kaiser first talked to me about his passion for orchids and their scents, I knew very little about these exotic and mysterious flowers.

However, he quickly fired my imagination, and now I am not only fascinated by the 'Flowers of kings' or the 'Queen of flowers', but I have developed a feel for the multiple facets linking our industry to the family of orchids.

I am excited by the way in which our researchers are taking their studies beyond the obvious, thus uncovering new horizons for perfumery, and am impressed with the way Roman Kaiser combines scientific thoroughness with artistic passion.

Givaudan-Roure can be proud to have researchers like Roman Kaiser in its team. When he joined the company in 1968, he brought with him the essential know-how which has finally led to this book on the Scent of Orchids: a chemical degree of the Technical College of Winterthur (Switzerland), competence in photography as a sophisticated hobby, and a deep understanding of botany. Indeed, Roman has been surrounded by flowers since childhood—his father was a professional gardener. So it was almost natural for him to develop a true passion for flower scents in the environment of Givaudan-Roure's renowned research and creative perfumery work; the fact that he received the prestigious Roche Research Prize in 1988 is just one of the fruits of this passion.

The variety of orchid scents is truly unique, and this inspired Roman Kaiser to dedicate a major part of his time to this flower species. The analysis of their headspace, a technology applied within Givaudan-Roure since the early seventies, has given us a deep insight into the essential characteristics of flower scents, and has uncovered a host of new chords which have very significantly contributed to our perfumers' creations.

I am convinced that, on reading this work on the Scent of Orchids, you will begin to share our passion, and like us become fascinated by the sensuality of a Laelia or a Cattleya.

Enjoy your fragrant reading,

Jean Amic
President and Chief Executive Officer
Givaudan-Roure

To this book

Since their discovery orchids have always been known for their marked beauty. As a result, they are grown extensively around the world in botanic gardens and by orchid enthusiasts, while thankfully they can still be found in most of their natural habitats.

Few of us realize that some of these unusual flowers, which we normally associate with a tropical paradise, can also possess an exotic and often penetrating aroma. This book takes the reader into this scented world of orchids. It is written by a unique scientist who has shown through the written word the love that he has for these fascinating flowers and their scents.

The book contains an introduction to the world of orchids, a description of the orchids found on five continents and the chemistry of their aromas. Rarely can such a text be found which captures the attention of those of us interested in floral aromas, pheromones, natural products, perfume creation and plant physiology, while still appealing to lay persons who share a common bond with the natural world. The Scent of Orchids by Roman Kaiser is unique in the way it has bridged the gap between the scientist and the lay person.

The book is fascinating to read. For example, it is amazing to learn that orchids are fitted with a 'biological clock'. Some orchids release their scent at night to attract their moth pollinators, while others, which are pollinated by bees, release their scents during the day. Kaiser discusses these facts and relates them to flower color in a section on pollination principles.

For the enjoyment of the lay person as well as the specialist, the book is filled with beautiful color prints of more than 170 orchids, so that the variation and exotic nature of many orchids are readily understood. To not overburden the lay reader with the detailed chemical composition with each of the aromas of the 165 orchid species examined, the author has cleverly placed the data in an appendix. This makes the book easier to read without compromising the scientific contribution of Kaiser's work.

Finally, here is a book of excellent value which will become the reference text on orchid aromas. In addition, it could easily become a prized coffee table book because it is easy to read while the rich colorful pictures are both interesting and beautiful.

Brian M. Lawrence
Editor-in-Chief
'Journal of Essential Oil Research'

Introduction

That most recent, yet most extensive, class of flowering plants, the orchid family, has spread across all the continents, gracing every vegetation zone with their astonishing vitality and variety. Over 25,000 species, subdivided into some 750 genera, are now known to exist, and no other plant family seems to offer such a wide variety of shapes, colours and, above all, scents. All the more surprising then, that one should continually encounter the mistaken belief that orchids are scentless. This erroneous view is doubtless due to the profusion of man-made hybrids, all too often scentless or else very weakly scented, which are currently offered by almost every florist.

From my own perspective, the picture is completely different. Over the last ten years I have documented approximately 2200 natural species, employing olfactory descriptions and photographs in the process. In about 250 cases, I investigated the orchids by 'headspace analysis', a technique which will be explained in the following pages, without destroying a single one of these remarkable flowers. Of these 2200 natural species, at least 50% may be classified as 'moderately' to 'strongly' scented, whilst only 15–20% proved, to my nose at least, to be scentless. Whether these particular orchids also appear scentless to their pollinators is quite another matter.

By dint of the special structure of the orchid flower, all orchid species, ignoring for the present the very rare phenomenon of self-pollination, are pollinated by animals—particularly insects and, less commonly, birds. Most of the generally known pollination principles apply to orchids, and one is not surprised, therefore, to discover an extraordinarily wide range of scents within the orchid family. In fact, I am personally convinced that the scent diagnoses of the other 4000 species of flowering plants recorded in my flower scent file all lie within the spectrum covered by the orchid scents.

The aim of this three-part book is to provide an initial impression of this huge variety and, as suggested by the title, to highlight the olfactory description and chemical analysis of the scents of the various species. Botanical and pollination aspects are also included, thus providing a reasonably comprehensive description of the phenomenon of 'flower scent'.

Part one concentrates primarily on explaining certain basic terms, methods and procedures that are subsequently employed in parts two and three. It starts with a few general remarks on the significance of the plant scent for humans and animals, goes on to describe the trapping and investigation of orchid scents and, finally, demonstrates how the huge variety of orchid scents can be understood as a reflection of the equally wide variety of pollination principles occurring in this family. The subject of the concluding section of part one is the 'Verbal description of scents', and this should help clarify the olfactory terms used in parts two and three.

In the second part of the book, the reader is invited to take part in a colourful, scent-enriched journey through the world of orchids, during which the scents and, to a certain extent, the overall appearance of some 160 species are described. In order to provide a sense of the plant geography involved, the presented species are summarized under three main distribution areas—the American tropics, the African tropics and the Indo-Australian tropics. Finally, we embark upon a short excursion to Central Europe to view some of the terrestrial species which, in my early youth, sparked my initial enthusiasm for orchids and, ultimately, led to this book. One of these distribution areas, at least, is presented in rather more detail through the selection of

Introduction

a relatively large number of neotropical orchids.

The third and final part of the book concentrates on the chemical analysis of a large number of the orchid scents presented and the structural elucidation and synthesis of new natural substances.

The astonishing variety of the orchid family and, more particularly, the high degree of variability even within a species is fascinating, and any investigation and documentation of the phenomenon of orchid scent must, of course, take this into account. There are species, naturally, that may be described as being constant both from the morphological and olfactory standpoints. With orchids, however, one must always guard against the assumption that the analytical results for one particular plant necessarily apply to the species as a whole. As we shall see—particularly in part one—the quantitative and, to some extent, the qualitative composition of an orchid scent depends not only on the time of day and the maturity of the flower, but also on the genetic circumstances of the individual plants. It should be stressed, therefore, that the olfactory investigations and chemical analyses of the various species were performed on the particular plants and flowers depicted in this book, and that—as a self-imposed restriction—non-destructive procedures were used. Nevertheless, given the sophisticated instrumental analytical methods currently available, this self-imposed 'orchid-friendly' approach restriction should have little adverse effect on the quality of the results.

As a result of the aforementioned relationship between the scent composition and temporal and genetic factors, partial or definite discrepancies may be apparent between the scent descriptions found (albeit infrequently) in the literature and those in this book. Verbal descriptions of orchid scents, presented in a wider context, are found particularly in the following books: 'The Scent of Flowers and Leaves' [1] by F.A. Hampton, 'The Fragrant Garden' [2] by L.B. Wilder, 'Duftende Pflanzen in Garten und Haus' (scented plants in the garden and indoors) [3] by F. Plenzat, and in articles by Müller [4], Richardson [5], Schnepper [6], Schwob [7] and Soule [8]. A recent publication by Nakamura, Tokuda and Omata [9] demonstrates the resurgence of interest in the traditional scent-awareness in respect of 'To-Yo-Ran' orchids (Japanese and Chinese *Cymbidium* species, *Neofinetia falcata* and *Dendrobium moniliforme*), an interest that also extends to the cultivation of 'Yo-Ran' orchids (i.e. all other species).

Although orchid scents very often appeal to our aesthetic sensibilities, their primary task is to attract animal pollinators, thereby contributing towards the preservation of the species. This discovery was first highlighted by Charles Darwin in his book 'The Various Contrivances by which Orchids are Fertilized by Insects' [10]. The corresponding classic of recent times—'Orchid Flowers: Their Pollination and Evolution' by L. van der Pijl and C.H. Dodson [11]—fully integrates flower scent within the discussion of the different pollination principles. The analytical investigation of orchid scents described in the literature has been undertaken mainly within this context, and it is not surprising, therefore, that the olfactory aspects play only a subordinate role in these works. A group of American scientists, including Dodson, Dressler, Williams and Whitten, concentrated especially on the investigation of the so-called orchid flower-euglossine bee relationship, whilst Kullenberg, Bergström, Nilsson, Borg-Karlson and others in Sweden dealt particularly with the pollination of *Ophrys, Cypripedium* and *Platanthera* species. Studies investigating these two subject areas which had been published by 1981 were reviewed by Williams [12,13] and Williams and Whitten [14], whilst more recent publications are cited individually in this book. For the sake of convenience, and (in accordance with the title of this book) to highlight the olfactory and analytical aspects of orchid scents in particular, the original botanical descriptions will not be cited during the discussion of individual species. Instead, a compilation of selected orchid publications is included in the appendix to assist those readers who wish to delve deeper into the subject.

Part One

**Introduction
to the World of Orchid Scents**

A few comments on the nature of orchids

From Confucius to Carolus Linnaeus

In European culture, itself based on ancient Mediterranean cultures, Theophrastus and Carolus Linnaeus are referred to as the fathers of the study of orchids. Theophrastus (372–287 B.C.), the Greek philosopher and a pupil of Aristotle, first used the term 'orchis' in his treatise 'Phytology' to describe the subterranean testicle-shaped tubers of the Mediterranean orchids. In the first century A.D., two orchids were called 'orchis'—in line with Theophrastus' description—by the Greek physician Dioscorides in his work 'Materia Medica', which described some 500 medicinal plants. Since the underground tubers of many of these 'orchis' species resembled human testes, Dioscorides—as a keen exponent of the doctrine of signature—attributed to them an aphrodisiac effect. Even today, salep—prepared from the dried 'orchis' tubers—is more widespread than one might perhaps expect in rural areas of Greece and Southwest Asia. In the mid-18th century, Carolus Linnaeus, in his seminal work entitled 'Species Plantarum', not only established the *Orchis* genus, but also extended the application of this name to the whole family.

In ancient China and Japan, orchids, and particularly the attractively scented *Cymbium* species, were highly prized long before we in Europe learned to appreciate them. Some 2500 years ago, Confucius lauded the beauty and scent of the flowers which he referred to as 'lan'. He compared their flowers to the perfect human being and their scent to the joys of friendship. This hymn of praise was probably intended for *Cymbidium ensifolium,* which has been grown in Japan and China for over 2000 years. *Cymbidium goeringii* could be the species that has been used, since ancient times, as a favourite motif in Chinese ink painting. It is mentioned in this book in respect of its extremely attractive scent (cf. analysis, p. 211).

Figure

1 *Orchis morio,* a species widespread throughout Europe, was probably a familiar sight in ancient times, as were certain other *Orchis* species. Their testicle-shaped tubers were used by representatives of the ancient doctrine of signature as an aphrodisiac.

A few comments on the nature of orchids

2

3   Photo H. P. Schumacher

In ancient Japanese and Chinese literature, frequent mention is made of orchids which now bear names such as *Cymbidium ensifolium, Cymbidium goeringii, Neofinetia falcata, Dendrobium moniliforme* and which are all characterized by intense, very pleasant scents. These 'To-Yo-Ran' orchids have retained their cultural importance down to the present day, and in Japan special orchid societies are dedicated to them.

On the American continent, orchids were first mentioned in the 'Codex Badianus', an Aztec plant book dating from 1552. The book describes the use of the vanilla pod as a flavouring for the traditional Aztec cocoa drink and as a 'perfume' in the preparation of 'tlilxochitl', a health lotion. To this day, *Vanilla planifolia* and *Vanilla pompona* remain the only orchids of direct economic importance.

Distribution and habitat

Over 25,000 known orchid species represent some 7–9% of all flowering plants, thereby constituting the family with the greatest number of species. With the exception of the Arctic, the Antarctic and hot, parched desert regions, orchids are distributed across all continents and zones, although 94–95% of species are concentrated in the Tropics. These tropical species exist primarily as epiphytes on trees, thus avoiding any competi-

Figures
2 A famous 'To-Yo-Ran' orchid: *Cymbidium virescens* (*C. goeringii,* Formosa) 'Setsuranhakka-Hakuun', excellence award from the fragrance judging at 'Japan Grand Prix International Orchid Festival '91'.
 Its exquisite perfume is reminiscent of lily of the valley and the fresh-floral aspect in the scent of fully ripe lemons.
3 *Vanilla planifolia* and, to a lesser extent, *Vanilla pompona,* are the only orchids grown as actual crops. Their flower scents, however, have little in common with the popular aroma of their pods. In the case of *V. pompona* (illustrated), the scent is 'aromatic-floral', with a characteristic hint of caraway.

tion with ground-based vegetation and remaining closer to the light. Since these epiphytic orchids are unable to obtain water from the ground, they are dependent, at least for part of the year, on high levels of humidity for their water. In order to survive dry periods, the shoots have developed into storage organs called 'pseudobulbs'. Many of the epiphytic orchids in the monsoon forests of Southeast Asia also shed their leaves during the dry season.

Some 5% of orchid species are located in temperate zones. In contrast with most tropical species, these orchids grow on the ground and are, therefore, known as terrestrial orchids. These species have developed underground storage organs (tuberous roots) which are capable of surviving the winter without those parts of the plant that are above the ground. Some orchid species have even ventured as far as the Arctic circle, the most famous example being *Calypso bulbosa,* which boasts a tropical-looking magenta flower during the short summer, subsequently surviving the Arctic winter unharmed in the frozen ground.

The orchid flora in each of the three main geobotanical distribution areas—the tropical regions of America, Africa and Indo-Australia—are highly diverse, although some genera can occur in two or even all three distribution areas at once.

Tropical America is extraordinarily rich in characteristic genera that only grow in this continent, including *Cattleya, Laelia, Epidendrum, Oncidium, Odontoglossum, Catasetum, Stanhopea, Masdevallia* and many others. Compared with the epiphytic species, terrestrial orchids play only a minor role. Overall, about 40% of all known species are native to the Neotropics.

Representatives of important epiphytic genera, including *Dendrobium* and *Phalaenopsis,* can be found throughout the Indo-Australian region, from the Asian continent via the Pacific islands, across to Australia. However, the numerous terrestrial species in Australia differ markedly from those of Asia. This huge geobotanical area is home to between 40% and 50% of all species. Relatively close connexions can be found, intriguingly, between the American and Indo-Australian orchids, and this is reflected in a certain congruity between the genera.

The African orchid flora, on the other hand, forms a completely uniform whole, accounting for only about 10% of species. Of particular interest is the fact that about a third of these African species are endemic to Madagascar. In contrast with the other two distribution areas, richly coloured, daytime-scented epiphytes are rare on the African continent, including Madagascar. Instead, there are very many white-flowering, night-scented species, including almost all the representatives of the genera *Aerangis* and *Angraecum,* which are given special consideration in this book. This difference has been ascribed to the historical development of orchids. Orchids are thought to have originated in the Asiatic-American supercontinent which existed before the present-day continents emerged. Africa is believed to have been colonized by orchids which had initially spread from the supercontinent; subsequently, however, it developed its own characteristic species.

The orchid species discussed in the second part of this book are summarized under these three main distribution areas. Within a given distribution area, the orchids are generally described in alphabetical order. For ready reference, a species index is also included in the appendix.

The flower
Orchids not only constitute the family of flowering plants with the greatest number of species, they also exceed all other plant groups in respect of variety of floral design. The stark contrast between the blandness of the flowerless plant and the almost indescribable splendour of shape and colour of the flowers—whose actual structure is no more complex than that of a tulip or lily—is an endless source of wonder. As with most monocotyledons, i.e. plants with a single embryo leaf, the structure of the flower is based on the number three, albeit with wholly characteristic variations.

A few comments on the nature of orchids

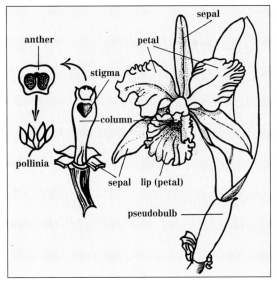

4

The six leaves of the flower are arranged in two concentric rings, or whorls, each made up of three leaves. The outer whorl consists of three sepals, the two lateral sepals being similar in shape and colour, and the median petal varying to a greater or lesser extent. Of the three petals that form the inner whorl, two are also similar, whilst the third possesses a different morphological structure and is known as the lip. In many species, the lip is the largest and most colourful part of the flower and serves as a suitable landing platform for the relevant pollinator. It can, however, be greatly diminished in size, as in the *Masdevallia* species, or be specially adapted to the pollinator by means of an extremely complex and sophisticated design, as seen in the euglossine orchids.

Since the ovary only swells after pollination, it is sometimes difficult to distinguish it from the stalk. During, or shortly before, flowering, the orchid is rotated through 180°, usually by the ovary. This process is known as resupination and causes the lip to move round to the lowest position, where it can be used as a landing platform. The principal modification, however, is experienced by the reproductive parts of the orchid flower, and this is a characteristic feature of the whole family. The male and female reproductive parts are combined in a uniform structure, known as the column, in the centre of the flower. In most orchids, the pollen grains stick together to form pollinia, differing in both number and shape, and located at the upper end of the column inside the anther. Directly beneath the anther is the female part of the column, the stigma, onto which the pollinium is rubbed off during pollination.

This hermaphroditic floral structure is the norm in orchids. Only a few species in the genera *Catasetum*, *Cycnoches* and *Mormodes* possess separate unisexual flowers which can exist on one and the same plant and yet flower at different times, thus making self-pollination practically impossible.

These few comments on the nature of orchids provide only a rough outline of these mysterious plants. They may tempt the reader to consult more specialist publications which not only discuss the morphological and ecological aspects but also provide fascinating descriptions of the discovery and scientific investigation of orchids. A relevant bibliography is included in the appendix.

Figures
4 Structure of an orchid flower
5 *Reseda odorata* (Resedaceae)

Plant scents

Relatively volatile substances, which in larger quantities would be toxic to the plants, are formed by complex metabolic processes in at least 30% of all the higher plant species. Because of their potential toxicity, they are stored as essential oils in special cells at the periphery of flowers, leaves, or even roots. Their volatility allows these essential oils, consisting of anything from a handful to over three hundred individual components, to be detected by the human or animal sense of smell. Humans generally experience them as a fragrant scent, although some plants can give off an offensive odour. Naturally, we cannot know how the pollinating animals experience them, although it may be assumed that they interpret all plant scents, regardless of how they are perceived by humans, as semiochemical messengers with a significant role to play in their own survival.

Flowers give off their scent without any visible external stimulus. They control the scent release by an internal biological clock which is particular to each species. Environmental factors, including light intensity, day-night cycle and temperature, act as triggering or inhibiting impulses, each affecting different species in varying degrees. Consequently, both the quantity and quality of the released scent often show a close correlation with time. As we shall see with the example of a whole group of orchids, certain species even appear to be scentless during the day but emit intense, and usually very attractive, scents after the onset of darkness. This phenomenon should be seen in the context of the biological significance of the flower scent. Its ultimate purpose, in conjunction with the shape and colouring of the flower, is to attract animal pollinators so as to ensure the preservation of the species.

Thus the scent of *Reseda odorata,* which humans also perceive as a very pleasant fragrance, is associated by its pollinator—the honey bee—with food in the form of nectar and pollen (Figure 5). The flowers of *Nicotiana alata,* which only release their scent during the night, are pollinated by a particular type of moth being guided, by the scent, to the nectar in the corolla tube (Figure 7, p. 19). The scent, colouring and shape of the flower of *Stapelia gettleffii* (Figure 8, p. 21) resemble carrion to female blowflies, which

5

Plant scents

6

therefore deposit their eggs on the flower, serving as pollinators at the same time. In both shape and scent, the flowers of *Ophrys insectifera* mimic the appearance and smell of the sexual attractant of the female sphecid wasp, thereby inducing the male to copulate with the flower, and in turn, resulting in pollination (Figure 6).

Certain species of flowering plants are even pollinated by vertebrates. For instance, the flowers which sprout directly from the trunk of the tropical American calebasse tree (*Crescentia cujete,* Bignoniaceae) (Figure 9) exude a fragrance which is reminiscent of fungi and leeks, thereby attracting a certain species of bat.

These five examples illustrate most of the pollination methods in which scent plays the major role.

Since the dawn of history, man, too, has been captivated by the scent—or perfume—of plants. The origin of the word 'perfume'—*per fumum,* literally 'through smoke'—almost suggests that scents were initially meant for the gods alone. In appealing to their mercy, man had to sacrifice to the gods his most precious possessions. These included fragrant resins, such as frankincense and myrrh, and, as civilization progressed, they increasingly took the place of sacrificial animals.

Thus, for example, the Epic of Gilgamesh, written in the 12th century B.C., and describing events in the 28th or 27th centuries B.C., describes how Ut-napishti, the father of all men, gave thanks for his rescue from the Flood by burning cedarwood and myrrh, whose sweet-smelling odour pleased the gods. The Egyptians were, after all, already expert in the manufacture and use of scented plant drugs and essential oils (including distillates). These substances were used not only in a ritual context, but also for medical and cosmetic purposes. The Egyptian Ebers Papyrus, written as early as 1600 B.C., contains the recipes of 100 such concoctions.

With the passage of time, man eventually learned how to extract the scents of many plants in highly concentrated form as essential oils. The technique of distillation may already have been known to the Indus Valley civilization as early as 3000 years B.C. But it was to be another 4000 years before this process was rediscovered by the Arabs. After further developments during the Middle Ages, it eventually assumed industrial forms in the 19th century. Figure 10 shows a typical example of the apparatus used by that time.

In this process, the scented plant material—e.g. flowering lavender branches, chips of

Figures
6 *Ophrys insectifera* (Orchidaceae)
7 *Nicotiana alata* Hort. (Solanaceae)
8 *Stapelia gettleffii* (Asclepiadaceae)
9 *Crescentia cujete* (Bignoniaceae)

Plant scents

7

8

Photo E. Scholz

9

Plant scents

sandalwood, or even scented roots like those of vetiver grass—is added to water and heated in the boiler. The fragrant volatiles evaporate with the steam and are subsequently returned to liquid form in the condenser. The essential oil thus obtained, which is insoluble in water, floats on the surface and can easily be separated off.

However, the more sensitive and valuable flower scents, such as those of jasmine, rose and tuberose, are usually extracted with a highly volatile solvent to produce, after various concentration stages, a 'flower absolute'. This method did not come into use until organic chemistry in the 19th century made available the appropriate solvents. In order to obtain, for example, 1 kg of rose scent in the form of essential oil or absolute by this classical method, 3–5 tonnes of *Rosa centifolia* flowers have to be picked by hand and processed. In the case of jasmine *(Jasminum grandiflorum)*, some 8 to 10 million individual flowers, weighing a total of 1 tonne, are required for 1 kg of absolute. It is not surprising, therefore, that both of these natural products, which are used in many perfumes, command as much as Sfr. 2000–8000 per kg. Nevertheless, among the 500 or so regularly used essential oils and absolutes, there are also those which can be obtained in much greater quantities with much less effort. Lavender oil, for example, only costs Sfr. 40–60 per kg, depending on the type and year.

With the development of organic chemistry towards the end of the 19th century, chemists working in the perfume industry began investigating the composition of these essential oils and related natural extracts. As a result of this (still continuing) research work, the perfumer now has at his disposal not only the 500 or so regularly used natural products for the preparation of his creations, but at least double this number of synthetic fragrance compounds. Nevertheless, the huge range of fascinating natural scents is still far from exhausted.

Many attractive flower scents, for example those of lily of the valley, honeysuckle, many tropical flowers and especially those of the orchids, cannot in fact be obtained as essential oils or absolutes of an acceptable quality, because important fragrance compounds are destroyed and highly volatile compounds are completely or partially lost during such physical processes, or else because the plant material produces artefacts which impair the scent. Moreover, the production costs for orchid extracts, for example, would be astronomical.

By the 1970s, methods of instrumental analysis—particularly capillary gas chromatography and mass spectrometry—had

Figure
10 Apparatus for extracting essential oils by steam distillation, 2nd half of the 19th century, Grasse, southern France.

reached such a high level of sensitivity that the scent given off by the living flower could be captured directly and the resulting samples subjected to analytical investigations. In the mid-1970s Givaudan-Roure thus implemented a long-term research program with the aim of investigating, and subsequently reconstituting synthetically, these fragile, highly-prized flower scents which were not available as natural products. Other companies within the fragrance industry probably started to address this problem at about the same time.

Before such investigations can be carried out, a sufficient quantity of the flower scent—if possible 10–50 millionths of a gram—must be available. This is approximately the amount of scent released into the air by a moderately fragrant flower over the course of an hour. This scent then has to be captured by some means or other. The most promising method, which can also easily be adapted for field experiments, involves the trapping of the scent given off by the flower with a small amount of a suitable adsorbent, such as activated charcoal or a porous polymer such as Poropak or Tenax. The method has since proved to be effective, and has been specially adapted over the last ten years to capture the orchid scents described in this book. This 'orchid-friendly' type of scent sample collection will now be described in more detail. A recently published review article by Bicchi and Joulain [15] provides a good overview of the literature describing such trapping techniques and their use in investigating plant scents. Moreover, a whole section of the book 'Perfumes: Art, Science and Technology' [16a] is devoted to the trapping and investigating of flower scents.

Trapping and investigation of orchid scents

The four examples shown below illustrate how the flower or the whole flower head (inflorescence) of the particular orchid species to be investigated can be enclosed by a glass vessel of suitable size and shape without damaging the plant or flower. This precautionary measure largely prevents further dilution of the scent by natural air circulation. The scented air surrounding the flower—known as the 'headspace'—is drawn through an adsorption trap—which is either attached to, or placed directly inside, the glass vessel—by means of a battery-operated pump over a period of 30 minutes to 2 hours (30–60 ml per min). Depending upon the species to be examined, the trap contains between 1 and 10 mg of very pure activated charcoal. Air and moisture pass unhindered through this microtrap, whilst the scent is adsorbed and accumulates. Depending on the intensity of scent

11

12

13

14

release, amounts ranging from 1 to 300 millionths of a gram are collected during the trapping period. A second adsorption trap, connected in series, ascertains whether all the scent components were actually adsorbed in the first trap.

Figures 11 *(Cattleya labiata)* and 12 *(Zygopetalum crinitum)* demonstrate the importance of using the correct glass vessel. Ideally, it should be adapted to the specific flower type so that on the one hand it enables the optimal scent concentration to be obtained and, on the other, keeps the risk of flower damage to an absolute minimum. However, for flowers with very complex shapes, e.g. *Angraecum sesquipedale* (Figure 13) which is subsequently de-scribed in greater detail, this could only be achieved at great expense. In such cases, it is more practical simply to isolate the flower from the environment as much as possible by a suitably shaped glass funnel. The adsorption trap is then centred as near as possible to the position where the scent release is at its

Figures
Trapping of orchid scents
11 *Cattleya labiata*
12 *Zygopetalum crinitum*
13 *Angraecum sesquipedale*
14 *Nigritella nigra*

maximum. For trips to remote regions, it is important to ensure that the experimental apparatus is not too heavy and cumbersome. With a little imagination, Erlenmeyer flasks, measuring cylinders or test-tubes can often be turned into fully functional 'headspace equipment'. Figure 14 shows how the scent of *Nigritella nigra* was captured in this way in the Swiss Alps at an altitude of 2500 metres.

The scene now changes from the natural location to the laboratory, where the trapped scent samples are investigated by the analytical methods described briefly below. Some of the specialist expressions used on the following pages may seem rather incomprehensible to the non-chemist. However, the reader should not be unduly concerned, since this will not affect his basic understanding of the main subject.

Back in the laboratory, these tiny scent samples of 1–300 µg must first be eluted from the adsorption filter using an equally small amount of a suitable solvent, e.g. 10–50 µl carbon disulfide. For complete recovery of the scents, which contain extremely polar compounds, e.g. free acids, further elution may be required with 10–30 µl ethanol. The scent solution thus obtained is of a suitable concentration to be investigated directly in respect of its qualitative and quantitative composition by the combination of capillary gas chromatography and mass spectrometry (GC/MS). Capillary gas chromatography makes it possible to separate complex scent samples into their individual components and to measure the quantitative proportions. Mass spectrometry plays an important role in elucidating the structure of each component. To help them carry out the GC/MS investigations more effectively, groups working in this specialist field are able to refer to comprehensive reference spectra files. These files make it much easier to identify known natural and synthetic products, thus freeing the researcher's time for structural elucidation of new natural products. If such a new product represents a major —or even the main—component of the scent trapping, it can be isolated by preparative capillary gas chromatography in sufficient quantities for subsequent structural elucidation by NMR spectroscopy. The corresponding spectra of the constituents can currently be measured with samples as small as 5 to 15 µg. Despite the enormous advances made in instrumental analytical methods, the small sample size of the preconcentrated scent still represents a limiting factor, since the structure of olfactorily important trace components cannot always be established. In order to obtain an interpretable mass spectrum, component quantities of the order of 1 ng must reach the mass spectrometer. Consequently, only components present to at least 0.1% in the scent trapping can be characterized by their mass spectra. However, some highly active scent constituents can be detected by the human nose at concentrations as low as a billionth of a gram (10^{-12} g) per litre. In order to learn about such trace components, at least as far as their olfactory properties are concerned, the most sensitive detector in this context—the human nose—is routinely used. To this end, part of the gas flow containing the scent molecules is passed through a sniffing tube fitted to the end of the capillary column. Whilst the chromatographic signal of a particular component is being recorded, the evaluating 'nose' (which is being used here instead of the mass spectrometer) notes the corresponding scent impressions and associations at the base of all the peaks on the gas chromatogram to produce an 'olfactogram' of the respective scent trapping. Figure 15 shows one such combination of a gas chromatogram and olfactogram for the scent of *Cattleya labiata*.

During scent reconstitution, the individual olfactory results for unidentified constituents can be used to match the corresponding scent notes to olfactorily related fragrance compounds. Furthermore, this information can help to determine whether trace components should be concentrated by preparative capillary gas chromatography to the point where they can be identified by their mass spectra, provided the sample size is sufficiently large.

In the olfactogram of *Cattleya labiata* (cf. Figure 15), two arrows indicate the location of two olfactorily important trace components which were below the detection limit of the

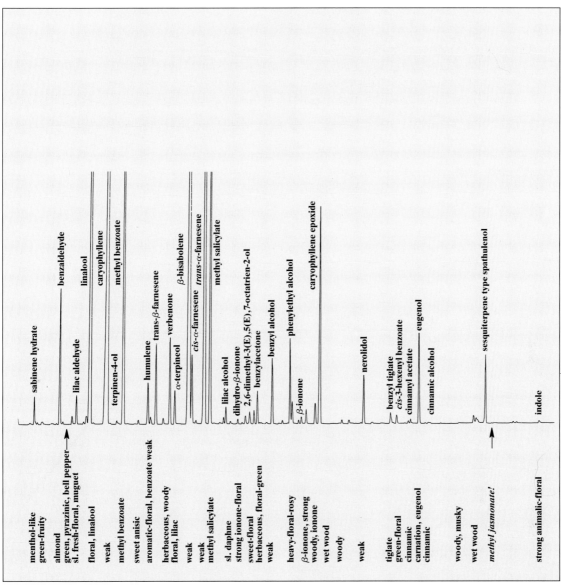

mass spectrometer. The first of the two components was eluted shortly after benzaldehyde and produced virtually no recorder signal. A strong odour of bell pepper was noted, however, during sniffing at the outlet of the capillary column. The chemist behind the nose realizes that this smell is represented chemically by 2-isobutyl-3-methoxy pyrazine. The second trace component, eluted just before indole, gave off the typical scent of methyl jasmonate and, as confirmed by a corresponding co-injection, methyl *cis*-(Z)-jasmonate is indeed eluted at this point.

The orchid scents described in this book were, without exception, trapped from the living, undamaged flower. It almost goes without saying, therefore, that the experimental conditions for sample collection were, as near as possible, adapted to the biological circumstances of the respective species.

As mentioned at the start of the book, the quality and quantity of the scent released from a flower is greatly dependent on its degree of maturity and on endogenous and exogenous factors. *Angraecum sesquipedale,* the well-

Figure

15 Gas chromatogram and olfactogram of the scent of *Cattleya labiata* (central section)

known night-scented orchid of Madagascar (Figure 20) vividly shows how the degree of maturity can affect the scent of a single flower [16]. The first chromatogram in Figure 16 refers to the scent trapped during the first night after flowering (inflorescence), the second to that collected during the following night.

Particularly significant differences can be seen in the relative contents of methyl benzoate, ethyl benzoate, phenylethyl alcohol and indole. Further details of the scent composition of this orchid are given on pages 30 and 197.

As already briefly mentioned, and subsequently detailed using the examples of the flower scents of *Odontoglossum constrictum,* *Citrus medica, Hoya carnosa, Stephanotis floribunda* [17–18], and a hybrid tea rose [16], the qualitative and quantitative aspects of the scent given off by a flower are often greatly dependent on the time of day. These biorhythmical aspects of scent release must, of course, be taken into account during the trapping of a scent sample if the corresponding analytical results are to be used in comparative investigations or are to serve as a basis for synthetic scent reconstitution. As we shall subsequently discover in greater detail, the orchid family is particularly rich in species with a high degree of time-dependency in their scent emanation. Thus, for example, *Epidendrum ciliare* (illustration, p. 90), which is almost

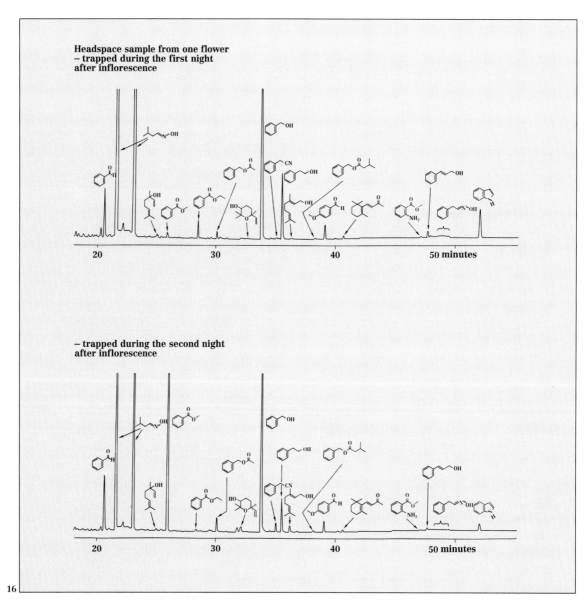

16

Trapping and investigation of orchid scents

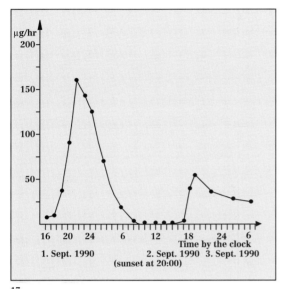

17

18

scentless during the day, shows a dramatic increase in scent production after sunset, reaching a peak at between 22:00 and 24:00 hr, and gradually subsiding during the early morning hours (cf. Figure 17). During the following night, the graph for scent emanation plots a similar course, though with diminished intensity. In order to obtain the results shown in Figure 17, a single flower was placed in a 150 ml glass vessel, specially adapted in shape and size, and the scented air was drawn through a charcoal trap (5 mg) at a suction rate of 35 ml per minute.

In concluding this discussion of the experimental and biological aspects of the trapping and analysis of flower scents, it should be pointed out that the composition of the scent of a defined species can vary from plant to plant to a considerable—or, in the case of actual chemotypes—even to an extreme degree. This applies in particular to the family of Orchidacea, in which species formation is by no means complete. It would not, therefore, require a specially trained nose to detect such differences in, for example, the highly varied scent of *Gymnadenia conopea* (Figure 18), a terrestrial species widespread throughout Europe. In a comparative investigation, we selected four representatives of *Gymnadenia conopea* at a site where there were some 50 flowering plants (Hostig, Uster, Switzerland, 30 June 1990). The scents were trapped between 9.00 and 11.30 hr and subsequently analysed by GC/MS. As can be seen from the compilation of the analytical results (Figure 19), the scent composition varies considerably from plant to plant. If the purely olfactory evaluation of the other plants is included, this particular location can be divided into two

Figures
16 Gas chromatograms of scent samples from *Angraecum sesquipedale*
17 *Epidendrum ciliare:* Emanation of scent during the course of two consecutive nights
18 *Gymnadenia conopea* (L.) R. Br., a terrestrial orchid widely distributed throughout Europe

main scent types. One group is characterized by a pronounced 'spicy-floral' scent, attributable primarily to eugenol and cinnamic alcohol, whereas the 'aromatic-floral' aspects of benzyl acetate predominate in the other group.

	1 %	2 %	3 %	4 %
Benzyl acetate	45.0	37.4	57.0	61.0
Benzyl benzoate	14.5	20.1	10.4	9.5
Methyl eugenol	5.7	6.2	5.9	6.1
Eugenol	2.6	9.5	0.3	0.2
Elemicine	3.3	3.1	5.2	5.4
Benzyl alcohol	2.6	2.5	1.4	1.5
Cinnamic alcohol	1.6	1.1	0.3	0.1
Phenylethyl alcohol	0.4	2.3	0.2	0.1
Phenylethyl acetate	0.4	0.5	1.4	1.3
(Z)-3-Hexenol	0.4	0.6	0.2	0.2

Figure
19 Variations in the scent composition of four *Gymnadenia conopea* plants

Scent and pollination principles

'Moth flowers'

As mentioned briefly in the introduction, a description of the enormous variety of scents, shapes and colours of orchids should also consider the equally wide variety of pollinating animals involved. At the one extreme, we have the group of moth-pollinated night-scented orchids. Most of these are coloured white, thereby providing additional guidance for the moth as it approaches the target. These orchids give off scents which are reminiscent of those emitted by jasmine, honeysuckle, tuberose, lilies, gardenia and the night-scented *Nicotiana* species—scents which are also released mainly during the twilight hours or at night. In the perfume trade, these scents are known as 'white-floral', and have been reproduced in a number of great perfumes.

An impressive example of co-evolution between a flower and its pollinator, these night-scented 'moth flowers' store much of their nectar at the base of a deep corolla tube or an elongated spur which can only be reached by the long proboscis of the moth. This adaptation of pollinator and flower is strikingly apparent in an orchid species native to Madagascar: *Angraecum sesquipedale* (Figure 20), which is completely scentless during the day, but which, with the onset of darkness, exudes an attractive and powerful scent until the following morning. This particular scent is reminiscent of a combination of scents from another night-scented plant, the South American night-scented sweet tobacco *(Nicotiana longifolia)* and the lily, and has overtones of the scent of *Gardenia*. This description applies to the fully developed flower, on its 3rd or 4th night after blossoming. During the first two nights, the scent is rather mixed, with a fairly marked indole note. This latter compound is found in the majority of night-scented flowers. At low concentrations it contributes a pleasant 'fresh-floral', at medium

Figure

20 *Angraecum sesquipedale* Thou. (comet orchid, Star-of-Madagascar orchid)

This, the most extraordinary of all *Angraecum* species, is native to Madagascar. The flowers, which have a spur up to 35 cm long, are scentless during the day but emit a pleasant 'white-floral', scent between dusk and dawn which is reminiscent of the lily and *Nicotiana longifolia*.

Scent and pollination principles

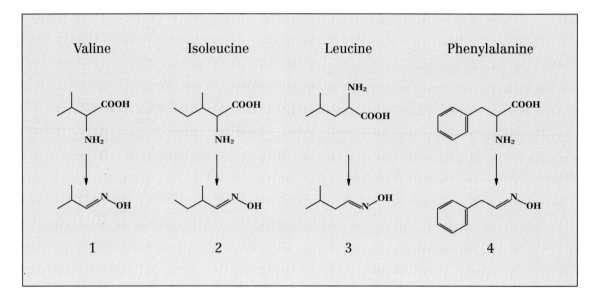

concentrations a 'narcotic-floral' and at high concentrations a rather aggressive naphthalinic or 'animalic-floral' aspect to the scent. Indole is particularly easy to recognize in the scent of orange flowers *(Citrus aurantium)*, for example.

According to the original description, the spur of *Angraecum sesquipedale* can reach the impressive length of 21–35 cm. When, in 1862, Charles Darwin described this magnificent flower, which varies in colour from ivory to brilliant white, he prophesied to botanists and biologists the appearance of an insect whose proboscis would be long enough to reach the nectar at the bottom of the extremely long spur. 40 years later, the corresponding pollinator—the moth *Xanthopan morgani*—was discovered in the warm forests of Madagascar. In Darwin's honour it received the annex *forma praedicta*. Taking the example of this and many other orchids, Darwin had specifically studied and identified the complex relationships involved in a process which, today, we refer to as 'co-evolution'. In this respect, he was far in advance of most of his contemporaries. As may be seen from the scent composition described on page 197 and the gas chromatogram depicted on page 26, the scent of *A. sesquipedale* contains large quantities of 3-methylbutyraldoxime (3), moderate amounts of phenylacetaldoxime (4) and small quantities of isobutyraldoxime (1) and 2-methylbutyraldoxime (2).

Over the last 10 years, we have identified these oximes, which may be considered to be metabolites of the corresponding amino acids [16], in about 30 different flower scents. In the majority of these cases, the respective flowers exhibit the typical syndrome of moth pollination.

Although no relevant studies have been reported in the literature to date, it is probable that these oximes possess semiochemical significance. They are not of overriding importance from the olfactory standpoint, but they do serve as 'scent modifiers', providing, in the case of *A. sesquipedale,* the characteristic complement to the 'basic scent', which is composed of benzyl alcohol, phenylethyl alcohol and cinnamic alcohol, plus derivatives of these alcohols, indole and certain other fragrance compounds (cf. analysis, p.197). By contrast, the nitriles and nitro-compounds generated from these oximes through dehydration and oxidation, which almost always occur as trace components in such flower scents, are strong-smelling compounds.

As will become apparent from a more detailed description of the *Angraecum* genus, these oximes may be considered, in the case of *A. eburneum,* as 'subspecies-characterizing' compounds.

Such night-scented, moth-pollinated orchids are especially common among the African genera, particularly among the *Angraecum, Aerangis* and other closely-related genera. This flower type is much less well represented in the American tropics, being restricted largely to the genera of *Brassavola* and *Epidendrum,* whilst in the Indo-Australian zone there are only isolated examples of these 'moth-pollinated' orchids spread over several genera. According to Pijl and Dodson [11], these orchids account for approx. 8% of all species. These particular plants and scents are described in greater detail in the relevant genus section.

'Fly flowers'
and related flower types

In dramatic contrast to the night-scented orchids just described, which are in most cases delightful to the human nose, are the putrid and fecal scents of those orchids which mimic carrion in both scent (stench) and colour (generally reddish-brown, greenish-brown or yellowish-brown) and thus attract carrion-feeding insects.

These are especially well represented in the very extensive *Cirrhopetalum* genus, which is spread across the whole of Southeast Asia. One unforgettable example is *C. robustum* (Figure 21), native to New Guinea, whose flowers emit a particularly penetrating stench resembling rotting flesh or overripe cheese. The South American genus *Masdevallia* also includes certain species, *M. caesia* for example (Figure 22), whose odour is almost as offensive.

Other fly and mosquito species in this pollinator group are more attracted by scents which, in man, would conjure up associations of crustacea and algae—the scent of the very small-blossomed *Malaxis monophyllos* for example (Figure 23) or of *Cirrhopetalum gracillium* (Figure 159), or else their odour might be reminiscent of wet dogs, or even of goats. To ensure the preservation of the species, orchids are able to comply even with these weird scent requirements.

As an illustration of this incredible talent for mimicry, we shall briefly consider the extraordinarily shaped flowers of *Dracula chestertonii,* an inhabitant of Colombia, which gives off a mushroom-like scent as a semiochemical signal, thereby attracting the female of a particular fungus-fly species (cf. Figure 24). Fungus-flies normally deposit their eggs on the nutritious receptacle of pileated fungi, where the larvae can subsequently feed off the fungal tissue. In common with certain other *Dracula* species, *Dracula chestertonii* seems to take advantage of this fungus-fly behaviour to ensure successful pollination. The unusually large lip imitates the cap of mushrooms occurring in the same biotope so precisely in terms of size, shape and smell that the fungus-fly becomes a victim of this mimicry, and deposits its eggs on the presumed mushroom cap. During this process, a pollinium inevitably attaches itself to the female insect by means of an adhesive disc (viscidium). On the insect's next 'mushroom visit', the pollinium is wiped onto the stigma of the second flower. Naturally, the hatched fungus-fly larvae are unable to find sufficient food on the lip and therefore perish. This biological principle has been described in detail by Vogel [72] with regard to various *Dracula* species as well as to flowering plants from other families. As explained in greater detail on page 219, the typical flavour compounds of mushrooms, such as 1-octen-3-ol and related compounds, are the main components of the scent of *D. chestertonii.*

'Bee flowers'

Between these two extreme groups—those orchids pollinated by moths and those pollinated by carrion-fly or related fly species—lies an enormous variety of orchids pollinated by bees, wasps and bumble bee species that cover a whole range of scents, from 'rosy-floral' and 'ionone-floral' to 'aromatic-floral' and 'spicy-floral' (see pp. 40–45 for an explanation of these terms). They include the familiar scents of lily of the valley, rose, sweet pea, primula, lime-blossom, mignonette, violet, hyacinth, narcissus and carnation, and any of the countless possible combinations of these scents. Occasionally, these scents are most elegantly enriched by woody, or even ambra and musk-like aspects, notes that are also used by the perfumer in the creation of floral perfumes. An extreme case is *Bollea coelestis* (cf. p. 51), a Colombian species with a strongly woody scent and containing large quantities

Scent and pollination principles

21

22

23

of caryophyllene and related compounds. In Pijl and Dodson's estimation [11], some 60% of orchid species are pollinated by hymenoptera.

As part of their adaptation to these bee species, the scent emanation of these orchids is dependent on daylight and warmth. As a further attractant, the respective flowers often possess intense colouring, with violet, blue, yellow, and occasionally white, serving as particularly attractive hues. In comparison with humans, the colour spectrum perceived by the honey bee—and, it is assumed, by most other bee species—is shifted by some 100 nm in the direction of the ultraviolet spectrum. Thus, whilst they are insensitive to red, they are still able to see wavelengths between 300 and 400 nm. The few predominantly red bee flowers almost always reflect ultraviolet light and can, therefore, still be detected by the pollinator. The shape of the *Cattleya* flower (Figure 25) is very typical of 'bee flowers' within the orchid family. The colour of the 3 sepals and the 2 petals, together with the (often intense) colouring of the lip, enhances the attractant effect of the scent, and the large, platform-like lip virtually invites the pollinating insect to land on it. Most of the 'bee flowers' in this family conform to this basic pattern.

But how does an insect find the nectar if, as with most of the orchids of this type, it is not freely available but concealed at the end of a corolla tube? In fact, the lip often possesses nectar guides, e.g. coloured lines that lead to the nectaries inside the flower. The nectar guides of white 'bee flowers' are often in the ultraviolet range and hence not visible to the human eye.

Within the hymenoptera pollinator group, the flower scent, the shape and colour of the flower are actually associated with food in the form of nectar in about 80% of cases. In the remaining 20% of bee, wasp and bumble bee species, the signals communicated by the respective flowers trigger a completely different behaviour pattern. The two most significant phenomena occurring within this group will now be briefly described.

Orchid-euglossine bee relationship

By far the strongest scented orchids, putting all other flower scents in the shade as far as intensity and quantity of the scent constituents is concerned, can be found within certain neotropical genera that are visited and pollinated exclusively by male euglossine bees. This orchid-euglossine bee relationship, which is particularly close in the genera of *Coryanthes, Gongora, Stanhopea* and *Catasetum,* has been specifically investigated by working groups headed by Dodson, Vogel, and Dressler. Comprehensive reviews of the literature up to 1982 have been published by Williams [12–13] and Williams and Whitten [14]. In contrast with most 'bee orchids', the species in these genera are not visited by their pollinators for their nectar—they do not, in fact, produce any—but are attractive on account of the large amounts of scent constituents which they produce.

The appropriate euglossine bee brushes off the scent droplets located on the flower surface with its hairy front legs and, from time to time, transfers the 'orchid perfume' to its back legs, from which it passes to a special container by capillary action. This whole collection process is concluded by a rather intriguing surprise for the bee (Figure 26), by which a pollinium is pressed onto the head or abdomen and subsequently wiped off again onto the sticky stigma of the flower of the same species next visited by the bee. This orchid-euglossine bee relationship is often highly

Figures

21 *Cirrhopetalum robustum* Rolfe grows in epiphytic fashion on tall trees in the rain forests of New Guinea at altitudes between 200 and 800 m above sea level.

22 *Masdevallia caesia* Roezl is endemic to Colombia, growing as a hanging epiphyte in rain forests at altitudes between 2000 and 2500 m. The flowers give off a strong, unpleasant smell resembling butyric acid, rotting flesh or stinkhorn.

23 *Malaxis monophyllos* (L.) Sw. is a rarely seen plant of central and northern Europe and northern Asia which grows in damp meadows and riverine forest. The flowers, which are 5 mm in diameter, emit a smell reminiscent of algae and crustacea.

Scent and pollination principles

24

Scent and pollination principles

specific, i.e. the flower scent and shape of an orchid species is, in many cases, designed to attract only a few, or even just one, of the 180 or so species of euglossine bees. In order to achieve this high degree of selectivity, the euglossine bee orchids have developed the most extraordinary floral shapes and scents (cf. examples, p. 73 seq.). But why is it just male euglossini that collect these orchid scents? All the theories currently proposed attribute to them a role in their own reproductive biology.

One theory states that these collected scent constituents are given off as a male aggregation pheromone, thereby causing males of the same species to swarm together and attract female euglossini by their glittering splendour. Another theory, and possibly the most convincing one, states that the orchid scent constituents are converted by the males into a sexual pheromone which then directly attracts the females. Although certain aspects require further clarification, it can still be safely assumed that these orchid scents are of crucial significance not only for the reproduction of the orchids, but also for that of the respective pollinator.

Ophrys flower-pollinator relationship

I shall conclude this brief discussion of 'bee orchids' by mentioning the equally fascinating *Ophrys* flower-pollinator relationship that ensures the continued existence of the 30 or so species in this European orchid genus. It has long since been known that *Ophrys* flowers, which can offer neither nectar nor utilizable pollen, are visited by certain male wasps and solitary bees. Some of the reasons for this have been discovered only in the past 30 years, notably by the Swedish working group headed by Kullenberg, and later by Bergström. A recently published review article by Borg-Karlson [19] provides a comprehensive overview of this subject.

It would appear that, as a consequence of the long-term co-evolution between flower and pollinator, the *Ophrys* flowers have actually managed to imitate, to a considerable extent, the sexual pheromone and appearance of the appropriate female bee species. Accordingly, it is only male hymenoptera species, never female ones, that are seen to visit *Ophrys* flowers. In the case of *Ophrys insectifera*, its scent—rather weak to the human

Figures
24 The flowers of *Dracula chestertonii* (Rchb. f.) Luer—an inhabitant of Colombia—emit a mushroom-like scent, the main component of which is 1-octen-3-ol. Moreover, the unusually large lip is deceptively similar, in both shape and colour, to the cap of fungi found in the same biotope.
25 The floral structure of *Cattleya* species is typical of 'bee flowers' in the orchid family. The rare *Cattleya rex*, from the Peruvian Andes, gives off a very pleasant spicy aromatic-floral scent.

Scent and pollination principles

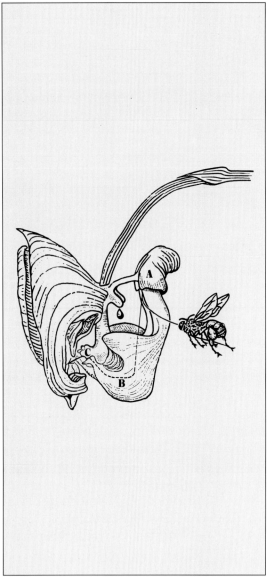

26

27

nose and reminiscent of lime-blossom accompanied by a slight aminic note—attracts the male fossorial wasp by fooling him into thinking that a receptive female is close at hand. If the wasp is within sight of the flower, he really believes that this is a female wasp. He is then sexually stimulated by special components in the scent and subsequently performs pseudocopulation with the flower, during which the two pollinia of the flower are pressed against the head of the fossorial wasp. After a while, the wasp senses that all is not quite right, and so he moves on to try his luck with the next flower. As he repeats the typical copulation movements, the pollinia that he has brought with him are rubbed off onto the stigma of the second flower, thus ensuring the cross-pollination of the *Ophrys* species. One big question remains, however: What rewards the fossorial wasp for all this frustration?

Although significant advances have been made over the last 20 years in the study of the *Ophrys*-pollinator relationship, we shall probably have to wait several more years yet before this rather curious phenomenon is fully understood. The latest research by the Swedish research group—and particularly that of Borg-Karlson [19 and references cited therein]—reveals a partial chemical correspondence between the *Ophrys* scent constituents and the secretion of the Dufour gland isolated from the females of the respective pollinators. Thus, the theory of chemical imitation of the female sexual pheromone propounded by Kullenberg does seem to be at least partially confirmed.

Butterfly flowers

The activity level of butterflies is dependent on daylight and warmth, and, in contrast with bees, butterflies are able to perceive red as a colour. Consequently, many 'butterfly flowers' are coloured red, pink or yellow. Although they can also be scented, visual contact appears to be of greater significance in attracting their pollinators.

Since butterflies are generally incapable, whilst hovering, of sucking nectar from relatively long spurs, the lip or lateral sepal of the typical butterfly flower has evolved into a suitable landing area. One particularly spectacular example is the South African terrestrial species *Disa uniflora* [11] (Figure 29), also known as 'The Pride of Table Mountain'. The two petals and the lip on this extraordinarily attractive orchid flower are greatly diminished in size, whereas the two lateral sepals form the most striking feature: a landing platform for potential pollinators.

The scent of *Disa uniflora*—which can only just be detected by the human nose and then only on particularly warm days—presents 'rosy-floral' and 'aromatic-floral' aspects. The typical butterfly flowers within the genus *Epidendrum*, *E. secundum* for example, are also only weakly-scented. The European terrestrial orchids include a number of representatives of this pollinator group that possess conspicuous 'spicy-floral' scents (see p. 40 seq. for an explanation of these terms). This applies particularly to the jewel of the European alps, *Nigritella nigra* (cf. p. 39, 176), whose 'spicy-floral' and aromatic scent is unmistakably redolent of vanilla and cocoa. This mountain orchid, normally coloured blackish-purple and, more rarely, salmon, pink, light yellow or whitish, is frequently visited by species of the Zygaenidae family [11], and, according to the author's own observations in Engadine (Switzerland), particularly by *Zygaena filipendulae* and *Procris statices* (Figure 30). A second European mountain orchid—*Traunsteinera globosa*—is also frequently visited by butterflies. Its scent is reminiscent of alpine hay, valerian and also, to a certain extent, of *Nigritella nigra*.

'Bird flowers'

In the tropical regions of both the Old and the New World, one continually encounters very brilliantly coloured orchids in various shades

28   Photo Prof. Kullenberg, Öland, Southern Sweden

Figures

26, 27 *Coryanthes* species induce pollination by causing the euglossine bee to slip whilst collecting the scent (A) into a trap (B) formed by the complex lip and partially filled with fluid. They can only leave the trap through a narrow opening (C) and are compelled to crawl under the stigma and through the pollinia. The various genera of these 'euglossine orchids' each possess their own version of this amazing pollination principle. *Coryanthes mastersiana* Lehm. (illustrated) is a Colombian orchid that gives off a dusty, leathery scent, containing as much as (or even more than) 90% of 2-N-methylamino benzaldehyde (cf. analysis, p. 210).

28 Pollination of *Ophrys insectifera* by a male fossorial wasp (*Argogorytes mystaceus* L.). In this case, the flower scent is not associated with food in the form of nectar. In fact, *Ophrys* flowers do not produce any nectar at all, but rather trigger the copulation instinct in the pollinator, causing it to approach and land on the lip. Copulation behaviour with the lip is also stimulated by its shape and surface texture.

Scent and pollination principles

Scent and pollination principles

of red and vermilion, which most closely resemble butterfly orchids but which, without exception, are scentless. In their relatively long corolla tubes, or even spurs, these flowers store nectar which is collected by various species of bird as food. This type of orchid does not need any scent, since the birds in question possess almost no sense of smell. Instead, they are attracted by brilliant colours, particularly in the red spectrum. This bird pollination syndrome is notably conspicuous in the humming-birds (Trochilidae) of the New World, which are specifically attracted, for example, to many species of the genera *Elleanthus, Cochlioda* and *Comparettia*. However, some representatives of the *Masdevallia, Epidendrum, Cattleya* and *Laelia* genera have also managed to adapt themselves to this syndrome. The Brazilian *Laelia harpophylla* (Figure 31) is a particularly striking example of the colouring of many humming-bird orchids.

In the Old World, sunbirds (Nectariniidae), in particular, are relatively common pollinators of *Dendrobium* species in Malaysia and New Guinea. A very impressive example is *Dendrobium lawesii* (Figure 32), growing in New Guinea at altitudes of 3000–3500 m. According to Pijl and Dodson [11], about 3% of orchid species are pollinated by birds.

31

32

30

Figures
29 *Disa uniflora*, a typical butterfly orchid in which scent plays a secondary role
30 *Nigritella nigra* with *Procris statices*
31 *Laelia harpophylla*, a typical 'humming-bird orchid'
32 *Dendrobium lawesii*, a typical 'sunbird orchid'

Verbal description of scents

The aim of trapping and analysing plant scents, orchid scents in our case, is to gain as clear an insight as possible into the composition of these scents and the structures of the individual scent information carriers. As already briefly mentioned (p. 24), it is still not possible, despite the highly sophisticated analytical systems currently available, to identify all the olfactorily important components in a scent. Even if the detection limit were lowered even further by a factor of ten, this statement would probably still be valid in 10 years. All the more important, then, to draw up a verbal description of the scent at the time of sampling, reproducible insofar as possible, so that this information is available in addition to the analytical data for purposes of scent reconstitution. In practice, many scents are not really amenable to chemical analysis, hence the need for an olfactory description, together with photographic documentation in the majority of cases. Of the 2200 orchid species documented in this way by the author over the past 10 years, chemical analysis was possible in some 250 cases. These scents, for the most part backed up by analytical data, now serve as internal standards which can be referred to in order to give the verbal description of related scents or scent aspects a greater degree of objectivity.

Publications dealing with the description and classification of scent impressions abound in the specialist literature. These are conveniently and attractively summarized in reviews by Amoore [20], Ohloff [21] and Thiboud [16b]. The methods described refer largely to defined fragrance compounds, often require a team of panelists and can only actually be used under laboratory conditions. The aim of all these methods is to render the scent description of an individual substance or a multi-component mixture amenable to database searching by employing suitable descriptors, or to give the chemist a molecular insight into the relationship between structure and smell. Over many years of describing plant and flower scents under field conditions, I have increasingly moved away from excessively narrow classification parameters and am now convinced that the widest possible degree of freedom must be allowed in order to obtain, at a later date, as accurate and comprehensive a picture as possible of these fleeting impressions. This freedom will of course involve, first and foremost, comparisons with previously investigated natural scents or, at least, with scents that are very familiar from the olfactory standpoint. Comparisons are also made with some 300–500 essential oils and over 1000 synthetic products at the perfumer's disposal. Impressions derived from the other senses, for example sweet, cold, green, fresh, warm, bright, soft or velvety, can also be employed comparatively, provided these rather subjective adjectives can be made more objective by further comparisons. For example, although 'green' is associated by most perfumers with freshly cut grass, and although it is known that this scent impression is produced by (Z)-3-hexenol and its derivatives, this adjective should not just be stated on its own, without further qualification. The scent of sliced cucumber is also experienced as 'green', but in this case, it is nonadienols and nonenols and their corresponding aldehydes that trigger the 'green' association. This is even truer of commonly used adjectives such as spicy, herbaceous, woody, fruity, balsamic and so on, which should always be rendered more objective by clear comparisons.

To employ the term 'flowery', without further qualification, to describe an orchid scent would be simply absurd. In fact, the term covers such a wide range of scents that its use can occasionally cause considerable difficulties when searching for associations with other defined scents if it is not properly divided into

Verbal description of scents

carefully defined subterms. There are various criteria which, depending on the location and scientific approach of the evaluator, can be employed to usefully divide such large scent concepts into subgroups. In my own case, on the basis of some 20 years' experience in scent diagnosis and analytical chemistry, it seemed appropriate to subdivide scents according to olfactory and chemical criteria. Subsequent experience has confirmed that flower scents can be subdivided into relatively well-defined groups using these criteria, with each group containing representatives of the most diverse botanical families. And in keeping with the ecological diversification of orchid flowers, so too the orchid scents extend across all these groups.

Four of these flower scent groups, which I believe are particularly significant, will now be described, together with details of the respective terms. The four groups are:
– Flower scents with a 'white-floral' image
– Flower scents with a 'rosy-floral' image
– Flower scents with an 'ionone-floral' image
– Flower scents with a 'spicy-floral' image.

For the sake of easy comprehension, the examples used by way of illustration have deliberately been chosen from among the very familiar 'perfumery-flower scents'. These specific terms can be applied unreservedly to the orchid family, and are highly practical in that they allow relatively complex olfactory relationships to be described in a few words.

This 'building-block' approach to flower scents also makes for an easier introduction to 'analytical sniffing', a technique briefly described at the end of this section.

Flower scents with a 'white-floral' image

The term 'white-floral' scent or 'white flower' perfume type has been used for a long time in perfumery, and incorporates the very pleasant, and interrelated, scent notes of jasmine, tuberose, orange flower, honeysuckle, etc. These flowers—which, as it happens, really are white in the vast majority of cases—release their scent mainly during the late evening or at night. Some of them, and notably the representatives of the orchids, are even scentless during the day, but are all the more strongly scented during the night (cf. pp. 27, 29 and 123). According to a rough estimate, they represent around 10% of scented flowering plants, and most are pollinated by moths which use the scent and the white colouring as pointers to the 'nectar-containing' flower. Since perfumes with this 'white-floral' image were all the rage during the 1970s and early 1980s, we investigated a whole series of flower scents in this olfactory group by the previously described method of 'headspace analysis'.

Figure
33 Flower scents with a 'white-floral' image

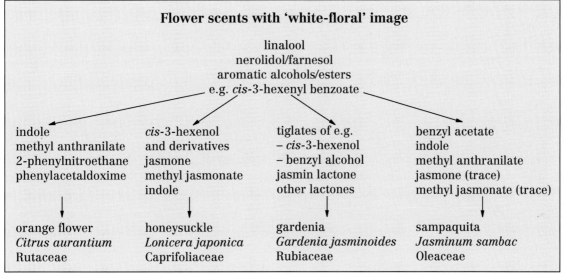

Verbal description of scents

Figure 33 illustrates how the olfactory affinity of such 'white flowers' is also reflected in their scent composition, using the examples of the scents of orange flowers, honeysuckle, gardenia and sampaquita *(Jasminum sambac),* the Far Eastern relative of the more familiar *Jasminum grandiflorum*.

The substances listed in Figure 33 demonstrate how the scents of these and many other 'white flowers' are largely based on the same underlying framework. Acyclic terpene alcohols such as linalool, nerolidol and farnesol, including the corresponding hydrocarbons, are accompanied by relatively simple aromatic alcohols (in particular benzyl alcohol and phenylethyl alcohol), the esters derived from them and esters of salicylic acid. Quantitative differences within this basic framework, together with species-specific compounds, make up the individual characterization of the respective floral scent. Amongst the orchids, such 'white-floral' scents are commonly found in the night-active species of the genera *Angraecum, Aerangis, Brassavola, Habenaria* and *Platanthera*. According to my own estimate, 'white-floral' aspects probably account for the scents of 5–8% of all orchid species.

Floral scents
with a 'rosy-floral' image

The term 'rosy-floral' scent relates particularly to the three famous scented roses that dominated the rosaria of Europe up until the beginning of the 19th century—*Rosa centifolia, Rosa damascena* and *Rosa gallica*. However, by no means all of the 150 or so natural species, let alone the thousands of hybrids derived from them, emanate scents that could be described as 'rosy-floral'. Particular mention should be made, in this context, of 'China roses', which have a completely different scent, and which reached Europe at the beginning of the 19th century. These were crossed with the aforementioned old European roses to produce tea roses and hybrid tea roses. The scents of these China roses—particularly crosses between *Rosa chinensis* and *R. gigantea*—very often contain large quantities of ionone compounds and 3,5-dimethoxy toluene, and have almost nothing in common with those of *R. centifolia* and *R. damascena*. Many tea roses and hybrid tea roses, direct descendants of the China roses, are also characterized by such carotenoid metabolites, and do not, therefore, belong to the group of 'rosy-floral' scents in this particular olfactory classification system, but rather to that of 'ionone-floral' scents. These 'rosy-floral' scents are found in the most diverse plant families. As with most of the 'white-floral' scents, many of them have a common basic composition. Taking the familiar scents of the cyclamen *(Cyclamen purpurascens)*, lily of the valley *(Convallaria majalis)*, sweet pea *(Lathyrus odoratus)* and *Rosa centifolia,*

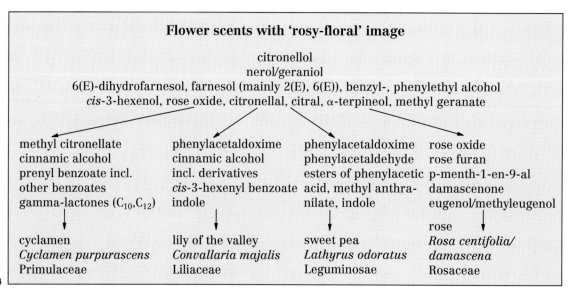

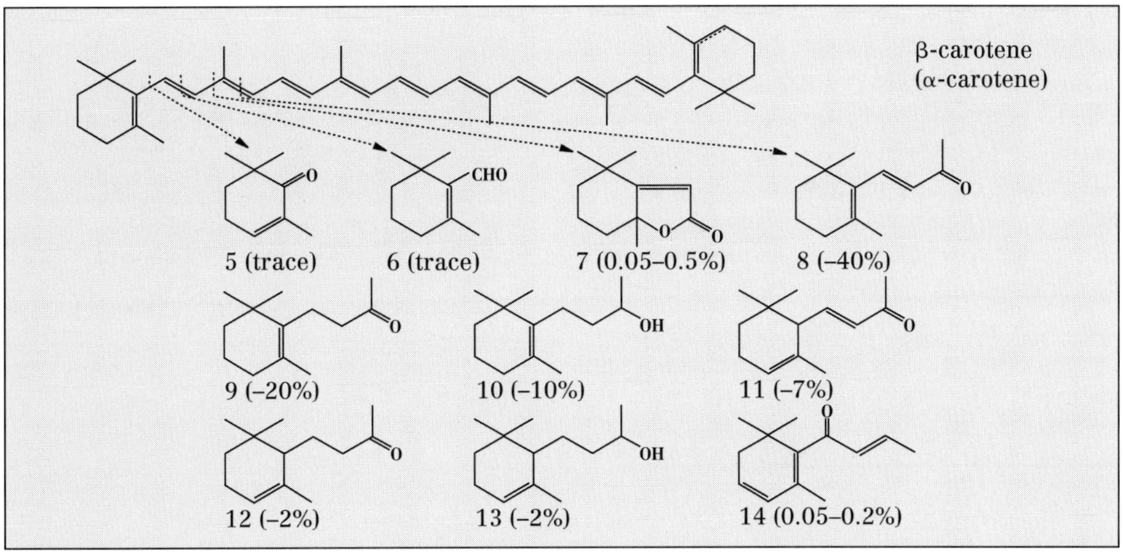
β-carotene
(α-carotene)
O
CHO
O
O
O
5 (trace)
6 (trace)
7 (0.05–0.5%)
8 (~40%)
O
OH
O
9 (~20%)
10 (~10%)
11 (~7%)
O
O
OH
12 (~2%)
13 (~2%)
14 (0.05–0.2%)

35

Figure 34 shows how species-specific compounds, together with quantitative differences, can again constitute the characterization of the individual species.

Whereas the 'white-floral' scents are mainly given off by night-active flowers, scent emanation from flowers with 'rosy-floral' scents appears in most cases to be triggered by sunlight and warmth. Interestingly, the orchid family itself contains a few notable exceptions to this rule, including the night-scented *Brassavola glauca* and *Brassavola digbyana* mentioned on page 59. Rather than giving off a 'white-floral' type scent, these flowers release a typical 'rosy-floral' scent. Pure 'rosy-floral' scents in the sense defined here are not very common amongst orchids. Many, however, border on to 'white-floral', 'ionone-floral' and 'spicy-floral' scents.

Flower scents with an 'ionone-floral' image

Flower scents with a so-called ionone-floral image form another significant group. Although this group incorporates very many scent facets and does not, therefore, appear to be any more homogeneous than the two preceding groups, the individual representatives have one highly characteristic feature in common: they are all rich in carotenoid degradation products, particularly in β-ionone and its derivatives. As might be expected, such flowers also contain the original carotenoids—particularly β-carotene—which give these flowers their colouring, often between orange-yellow and yellow-brown. However, flowers coloured by carotenoids only release their 'ionone-floral' scents if, at the same time, they possess the corresponding 'cleaving enzymes'. Given the prevalence of β-carotene in such flowers, β-ionone (8) is the predominant scent component in most instances (Figure 35), almost always accompanied by smaller quantities of dihydroactinidiolide (7). Dihydro-β-ionone (9) and the corresponding alcohol (10) are also major components, whilst α-ionone (11) and its derivatives (12 and13), together with damascenone (14), are normally only present in small quantities. The percentages shown in Figure 35 show the approximate upper limits for the amounts of these compounds in 'ionone-floral' scents.

Included in this olfactory group are the flower scents of the familiar freesia (Iridaceae), *Michelia champaca* (Magnoliaceae), a species growing in the Philippines and Indonesia, the Chinese species *Osmanthus fragrans* (Oleaceae) and the Western Australian species *Boronia megastigma* (Rutaceae). These scents are of such intensity and richness, and so unmistakeable in char-

Figures
34 Flower scents with 'rosy-floral' image
35 Degradation products of carotenoids commonly found in flower scents

Verbal description of scents

acter, that they are ideal candidates for illustrating the variety of 'ionone-floral' scents.

In orchid scents in particular, ionones are often accompanied by large quantities of aromatic compounds which also add a characteristic note to the scent. Above all, these include benzyl, phenylethyl and cinnamic alcohol as well as the derivative esters and corresponding aldehydes. Apart from references to other scents in this book, the term 'aromatic ionone-floral' is also used to describe such scents.

Of the scents of 250 orchid species investigated to date, β-ionone has been detected in 30 representatives, and I never cease to be amazed by the endless variety of guises in which this well-known perfumer's fragrance compound appears. This capacity of β-ionone to form homogeneous scent chords probably explains why its purely olfactory recognition can sometimes pose problems for the less experienced evaluator.

Flower scents
with a 'spicy-floral' image

This final group of flower scents appears to be rather heterogeneous since its representatives are very often on the borderlines of the other aforementioned groups of flower scents. One feature common to all of them, however, is a characteristic complex of phenolic compounds, frequently composed of p-cresol (15), vinyl guaiacol (16), chavicol (17), eugenol (18), isoeugenol (19) and vanilline (20) (cf. Figure 36).

These phenolic compounds are very distinctly expressed in the carnation *(Dianthus caryophyllus)*, which gives off the prototype, as it were, of a 'spicy-floral' scent. Amongst the orchids, it is the extraordinarily attractive *Masdevallia glandulosa* (illustration, p. 102), that possesses the most typical 'spicy-floral' scent.

Certain representatives of this complex of phenolic compounds are extremely widespread in orchid scents. Eugenol (18) and vanilline (20), in particular, have been identified in 49 and 47 cases, respectively, out of 250 species. Because of its low volatility, however, vanilline is seldom observed in concentra-

36

tions exceeding 1% in the headspace samples used for chemical analysis. In several cases, the phenolic compounds 15–20 also occur in fairly large quantities in the scent of night-active orchids whose basic image is of the 'white-floral' image. A typical example is *Angraecum bosseri,* illustrated on page 133. The term 'spicy white-floral' is used to describe these scents. As with the ionones, the spicy complex consisting of compounds 15 to 20 can also be accompanied by considerable quantities of aromatic compounds (cf. p. 43), in which case the designation 'aromatic spicy-floral' may be used. Scent transitions are also encountered, producing the designations 'rosy spicy-floral' or 'ionone-like spicy-floral'.

Example of a scent description

Taking the example of the scent of *Cattleya labiata,* whose chromatogram and olfactogram are illustrated on page 25 (flower illustration, p. 62), I should briefly like to explain the process by which a scent description is arrived at.

On the corresponding file card four completely separate entries have been recorded since 1983. Multiple diagnoses are required for a flower scent, on the one hand to monitor the sensitivity of the nose as a living analytical instrument and, on the other, to obtain sufficient data concerning biological aspects of the scent, such as the correlation between the scent and the flower's maturity or the time of day.

The entry for 4 November 1988, 11.30 hr, St. Gallen Botanical Gardens, reads as follows: "*Cattleya labiata,* type: *vera.* Pronounced 'aromatic spicy-floral' with a velvety aspect, but nevertheless with a transparent overall impression; very slight green-pyrazinic hint of hyacinth.

The 'aromatic floral' part is reminiscent of tiaré *(Gardenia taitensis):* methyl benzoate, methyl salicylate and homologues. After subtraction of this tiaré aspect, a 'rosy-floral' note is discernible, reminiscent of linalool, citronellol and phenylethyl alcohol. The 'spicy' part of the 'spicy-floral' aspect is mainly reminiscent of eugenol, followed by isoeugenol, vinyl guaiacol and a touch of vanilline, possibly also cinnamic alcohol. 'Velvety aspect': common in unifoliate cattleyas, reminiscent of sandal compounds, caryophyllene epoxide and, very slightly, of macrolides."

This description shows how the scent is initially perceived as an integral whole; only at a later time can the individual scent aspects be experienced by means of suitable methods. At first glance, these 'suitable methods' might give the impression of being completely individual, but they are ultimately based on a technique that can be described as 'analytical sniffing'. Once the qualitative content of one scent aspect, usually the dominant one, has been registered and has received sufficient attention, this scent—the first to enter the sniffer's consciousness—is 'subtracted' as far as possible to allow new scent notes to come to the fore. If associations with previously investigated scents are triggered during this process, the scent diagnosis becomes much simpler, since new points of reference for 'analytical sniffing' are thereby given. It should be emphasized that this method of scent description requires a great deal of time and concentration and cannot, therefore, always be used to its fullest extent.

Figure
36 Phenolic compounds important to 'spicy-floral' scents

Part Two

Interdisciplinary Discussion of Orchid Scents

Orchids of the American tropics

Acacallis Lindl.

Having been separated from the closely related genus *Aganisia,*

Acacallis cyanea

is currently the only species in this genus. This magnificent orchid was discovered in 1851 at the confluence of the Rio Negro and the Amazon, but was not cultivated until some 30 years later and remains a rarity to this day. Its native habitat, the upper Amazon and its tributaries, is one of the hottest and most humid regions of the world. This epiphytic orchid grows near the ground in marshy lowland forests or in the immediate vicinity of rivers. Its pencil-thin rhizome can extend for several metres, and is punctuated every few centimetres by pseudobulbs.

The flowers, growing up to 6 cm in diameter, are bluish-violet with a delicate hint of pink on the outside, and white on the inside. The golden bronze and reddish-violet lip forms an extremely attractive colour contrast. Its allotted genus name '*Acacallis*'—one of Apollo's beloved nymphs—is a tribute to the ravishing beauty of this flower. *A. cyanea* is probably the most famous example of the rare 'blue orchids' which only appear in a limited number of genera. Its rather weak, though highly memorable, scent has a metallic-cool aspect and contains up to 80% myrcene, accompanied by smaller amounts of eucalyptol, caryophyllene, ocimene and p-cresol. The visual and olfactory effect of this orchid contrasts strongly with the luxuriant vegetation of its preferred biotope.

Acineta Lindl.

This genus, comprising some 12 species of epiphytic orchids is found over an area extending from southern Mexico to Venezuela and Ecuador. The pseudobulbs, normally very substantial, carry between two and four applegreen leaves, which are long and often plicated. The raceme, which emerges from the base of these pseudobulbs, is clustered

Figure
37 *Acacallis cyanea* Lindl.

Orchids of the American tropics

38   Photo G. Gerlach

and pendulous, reaching up to 1 m in length and densely packed with large, strangely coloured and intensely scented flowers.

One of the most famous representatives of this genus, so rarely found in collections, is

Acineta superba,

often known as *A. humboldtii*. This splendid orchid was, in fact, discovered at the start of the 19th century by A. von Humboldt and A. Bonpland, who referred to it as *Anguloa*. The flowers, up to 8 cm in diameter, vary greatly in their colouring, from light yellow to reddish-brown, and are covered with red to brownish-purple spots. The scent can also vary considerably from plant to plant, particularly as regards the dominance of its two main aspects, on the one hand a spicy-cinnamic note and, on the other, a characteristic waxy, hesperidic aspect. These olfactory notes are principally due to the presence of methyl cinnamate and (Z)-α-*trans*-bergamotol (cf. analysis, p. 188). The latter compound accounts for about 44% of the analysed sample and, until now, was only recognized as a minor constituent of East Indian sandalwood oil [22].

Anguloa Ruiz et Pavon

The orchids of this genus, consisting of at least ten species, grow epiphytically or terrestrially on moss-covered rocks in the Andes, from Colombia to Peru, at altitudes between 1500 and 2200 m. Several unifloral scapes emerge from the base of the substantial pseudobulbs, which bear from two to four plicated leaves. The structure of the flower is reminiscent of a tulip. Its parts are all very fleshy, particularly the column, concealed within the flower, and the lip. The lip itself is flexible enough to swing back and forth between the column and the almost fully-united sepals. This explains why the *Anguloa* species are often known as 'cradle orchids'. Here, too, the unusual floral structure is designed to assist pollination, a service performed by male euglossine bees [11].

39

Anguloa clowesii

The species most commonly found in collections is probably *Anguloa clowesii*, a native orchid of Venezuela and Colombia. The large yellow flowers, at first glance suggestive of tulips, give off a familiar scent whose name does not readily spring to mind. It has an aromatic, warm-herbaceous character, with a refreshing eucalyptus aspect. In fact, a well-known fragrance compound—hydroquinone dimethyl ether—accounts for up to 90% and more of the scent, accompanied by small amounts of eucalyptol (cf. analysis, p. 197). *Anguloa uniflora* possesses a similar scent composition, since it also contains up to 85% of hydroquinone dimethyl ether. This compound is one of the most widespread of orchid constituents, occurring in 46 of the 250 species investigated. However, the two *Anguloa* species just mentioned represent extreme examples as regards the percentage of this substance contained in the scent.

Bollea and other 'fan-type orchids'

The rainy regions of the American tropics are home to a group of related orchids which, instead of developing pseudobulbs for purposes of water storage, have leaves which are arranged in a striking fan-shaped design. These bulbless 'fan-type orchids' are all interrelated and include certain species from the following genera: *Bollea, Chondrorhyncha, Cochleanthes, Huntleya, Kefersteinia* and *Pescatorea*. It would not be an exaggeration to state that this group contains many spectacular floral shapes and colours, and it is not remotely surprising to discover that many orchid lovers specialize in these particular plants, some of which are very rare and notoriously difficult to cultivate. No less fascinating are the scents of these 'fan-type orchids', ranging from 'rosy-floral' through woody, to yeasty. My own impression is that this group conceals the largest number of new scent constituents.

40

Bollea Rchb.f.

The genus *Bollea* is very closely related to the genera *Pescatorea* and *Huntleya* discussed below, and species of these genera often grow together in the same rainforest biotopes of Brazil, Colombia and Ecuador. The commonest representative is

Bollea coelestis,

an epiphytic orchid that grows at altitudes of 1700–2000 m, low down on the trunks of giant primeval forest trees.

The bluish-violet coloured flowers grow to 8–12 cm in diameter and give off a highly conspicuous woody, balsamic scent that I have never encountered in any other plant. Under certain circumstances, a degree of patience is required before this olfactory experience can be savoured, since the flowers only begin to release their scent two to three days after blooming, and the intensity of the scent falls sharply again after a further two days. This is yet another illustration of the need to evaluate the same plant at different times in order to obtain a reasonably definitive scent diagnosis.

Figures

38 *Acineta superba* (H.B.K.) Rchb.f.
39 *Anguloa clowesii* Lindl.
40 *Bollea coelestis* (Rchb.f.) Rchb.f.

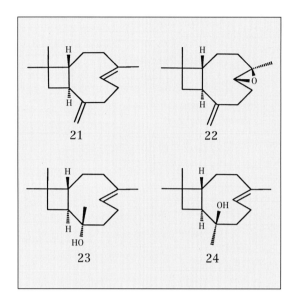

As shown by the analysis described on page 197, the main component of this highly unusual scent is the commonly occurring natural substance caryophyllene (21), accompanied by caryophyllene epoxide (22). The next largest component, and one particularly responsible for the special woody note, is caryophyll-5-en-2α-ol (23). This substance had not, hitherto, been recognized either as a natural or a synthetic product, nor have we managed to identify it in any other orchid scent. By contrast, the corresponding epimeric alcohol (24), which also had not been recognized as a natural substance until now, is contained in the scent of *Gongora cassidea* in trace amounts. We described this alcohol, easily prepared synthetically (24), some 15 years ago in connexion with the structural elucidation of caryophyllan-2,6-β-oxide isolated from verbena oil [23].

Chondrorhyncha Lindl.

The relatively few species in this genus, restricted to the tropics of Central and South America, are very closely related to those of the genus *Cochleanthes* described below. Certain of these species are characterized by bizarrely shaped lips, for example

Chondrorhyncha chestertonii,

of Colombia, which Hawkes [24] describes as being strongly scented.

Personally speaking, I have yet to encounter a strongly scented flower of this species, having merely been aware of a slight 'green note' reminiscent of violet leaves. Accordingly, the most northerly representative of this genus, the tropical Mexican species

Chondrorhyncha lendyana

is all the more strongly scented, though not especially appealing since the scent is evocative of a bug's defensive secretions, accompanied by a hint of yeast. This is attributable mainly to the two isomers of 2,4-decadienal and (Z)-4-decenol (cf. analysis,

p. 207). As will be discussed subsequently in greater detail (cf. p. 119), these isomers of 2,4-decadienal and the corresponding derivatives are relatively widespread in orchid scents.

Cochleanthes Raf.

More stimulating and varied scents than those of the *Chondrorhyncha* genus are produced by the closely related genus *Cochleanthes,* which encompasses some 14 species within the same distribution area. The appearance of

Cochleanthes aromatica,

a native of Costa Rica and Panama is particularly eyecatching, with flowers in shades of green, white and bluish-violet.

Equally striking is its powerful scent, which suggests an accord made up of rose, narcissus and verbena. Although the individual aspects can vary in intensity accord-

Figures
41 *Chondrorhyncha chestertonii* Rchb.f.
42 *Chondrorhyncha lendyana* Rchb.f.
43 *Cochleanthes aromatica* (Rchb.f.) R.E. Schultes et Garay

Orchids of the American tropics

ing to the plant and the age of the flower, geraniol, geranyl acetate, neral and geranial invariably form the main components, and are accompanied by olfactorily significant quantities of eugenol, vanilline and indole (cf. analysis, p. 208). Indole, in particular, can vary considerably, reaching concentrations as high as 5% in certain clones. In such cases, one flower is sufficient to fill a whole room with this unmistakable scent.

No less attractive is

Cochleanthes discolor,

which grows on moss-covered boughs from Honduras to Venezuela. Whilst the light green petals are barely discernible, the dark violet lip with its brilliant yellow callus on the inside makes an eyecatching spectacle. As one approaches these flowers, the nose, too, is captivated, although the sniffer may have to wait awhile before being able to associate its scent with gurjun balsam, wood and pepper.

This highly unusual scent is attributable to by certain extremely labile sesquiterpenes, particularly germacrene A (25), and a ketone derivative with a molecular weight of 220 whose structure, according to the NMR data, appears compatible with that of germacra-1-(10),11-dien-5-one (26) (cf. analysis p. 208). However, this structure has yet to be confirmed by synthesis.

45   Photo Ch. Weymuth

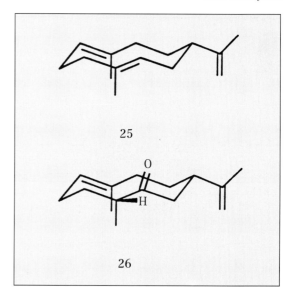

46

The third representative of this genus,

Cochleanthes marginata,

neatly illustrates how the huge variety of scents within the orchid family can be reflected in just one small genus.

Whilst *C. aromatica* presents a fresh floral bouquet and *C. discolor* emits a strange and surprising balsamic, woody note, the voluptuously arrayed flowers of *C. marginata* give off a very soft aromatic-spicy scent, containing up to 70% methyl cinnamate (cf. analysis, p. 209).

Huntleya Lindl.

Even smaller, but no less fascinating, is the fourth genus of the 'fan-type orchids', that of *Huntleya*, which encompasses just five species that are very rarely found in orchid collections. Being bulbless orchids, they all love hot, humid lowland rainforests, and this

Figures
44 *Cochleanthes discolor* (Lindl.)
R.E. Schultes et Garay
45 *Cochleanthes marginata* (Rchb.f.)
R.E. Schultes et Garay
46 *Huntleya meleagris* Lindl.

Orchids of the American tropics

47

48

probably explains why so few plants survived the long journey to Europe before the advent of rapid air travel. The specialist literature includes many reports describing how most of these extremely rare plants, collected under very adverse conditions, dried up or rotted away in some ship's hold.

Undoubtedly the most famous species of this genus, and possibly the most highly prized of the 'fan-type orchids', is the rare 'Star of the Rainforest'

Huntleya meleagris,

which is found growing from Colombia to Guyana. The wax-like flowers of this plant vary greatly in size and colour. A particularly sought after specimen is illustrated on page 55, possessing a scent at least 10 times more powerful than that of a relatively common variant with a white centre. This variant probably accounts for the fact that *H. meleagris* is sometimes described as scentless.

But if the right plant is sniffed at the right time, one is soon convinced of the opposite, and is completely captivated by this unique, unmistakable plant. The difficulty lies in recalling where this familiar sweet, woody-camphoraceous scent note was last experienced. The experienced sniffer will cast his mind back to the 1970s, to the era of sandalwood chemistry. Of course: the substance in question is *trans*-verbenol, a by-product of campholenic aldehyde, which is so crucial to the synthesis of various sandal fragrance compounds. *trans*-Verbenol is the predominant component but is accompanied by a whole series of related compounds (cf. analysis, p. 226).

Quite different in its general appearance and scent is

Huntleya heteroclita,

an epiphytic Peruvian orchid first described in 1944. Its 'rosy-floral' scent is heavily dominated by geraniol and nerol and corresponding acetates. The scent of this rare species also contains large quantities of the new natural substance 2(3)-epoxy-geranyl acetate (cf. analysis, p. 226).

49

Pescatorea Rchb.f.

The last genus of these 'fan-type orchids' to be touched upon is *Pescatorea,* a genus encompassing some 14 species that grow from Costa Rica to Colombia. As with the *Huntleya* genus, the relatively short pedicels (flower-bearing stalks) emerge from the base of the fan, bearing very large, eye-catching, and consistently fragrant flowers.

On the other hand, the predominant scent of

Pescatorea cerina,

a species growing in the humid forests of Costa Rica and Panama, is a very characteristic green-aldehydic aspect, reminiscent of yeast. This highly distinctive note is encountered as a 'shadow' in a relatively large number of orchid scents, producing some very interesting effects at certain dilutions. Analysis (cf. p. 239) reveals that the extremely strong-smelling substance (Z)-4-decenal is the main contributor, accounting for up to 13% of the volatiles.

As with *Chondrorhyncha lendyana* (p. 52), the rather unpleasant aspect of this floral scent is attributable to the two isomers of 2,4-decadienal. In sharp contrast as regards appearance and scent is

Pescatorea dayana,

a species found in northwest Colombia which occurs in a number of differently coloured variants.

This plant, illustrated above, has a pleasant 'aromatic rosy-floral' scent, which is dominated by a combination of benzyl acetate, geraniol and cinnamic alcohol, including derivatives (cf. analysis, p. 239).

Brassavola R.B.

This small neotropical genus incorporates about 15 species extending from Mexico, via Central America and the West Indies to Brazil. However, in contrast with the 'fan-type orchids' just described, they often prefer those regions where temperature and humidity fluctuate between extremes. Accordingly, they have adapted to these climatic conditions by developing a succulent growth form. The shoots have thickened to form robust pseudobulbs and, like most cylindrical or rounded leaves, possess a leathery texture. A greyish wax layer, often found covering the leaves, provides additional protection against transpiration. Most striking, however, are the flowers, which are coloured white to white-greenish in all the species, and also the strong

Figures
47 *Huntleya heteroclita* C. Schweinf.
48 *Pescatorea cerina* (Lindl.) Rchb.f.
49 *Pescatorea dayana* Rchb.f.

scent, which is only released during the night. *Brassavola* is the only neotropical genus in which all species show the typical syndrome of moth pollination. A paper dealing with the floral scent constituents of *Brassavola* species was published by Williams [25] in 1981. The most widespread species is

Brassavola nodosa,

growing from Mexico to Venezuela in coastal regions at altitudes up to 500 m. As a rule, this very succulent species grows epiphytically, occasionally on extreme stocks such as columnar cacti, coconut palms or tamarind trees. Many specimens have advanced as far as the coastal mangrove forests, even being washed by sea spray in some cases, or else grow lithophytically on precipitous coastal rock faces.

Although this orchid, mentioned even by Carolus Linnaeus, often grows in profusion, its distribution almost amounting to that of a 'weed', it is still held in high regard in Central America where it is known by the name of 'Dame de la Noche' or 'Alluring Lady of the Night'. This was doubtless inspired by the conspicuous olfactory impact of its scent, which is released after dusk and varies from a pure 'white-floral' to a 'rosy-floral' character, depending on the particular variety or subspecies.

The 'white-floral' aspect predominates in the small-flowered specimens so often encountered in orchid collections. It is based primarily on linalool (27), benzoates, salicylates and large amounts of nerolidol (28). The illustrated specimen, originating from Mexico and possessing relatively large flowers and a somewhat broader lip with rather flaccid petals and sepals, is characterized by a particularly diffusive scent, representing a combination of 'white-floral' and 'rosy-floral' scent notes. The latter aspect is mainly due to a high proportion of geraniol (29). The interplay of the two main components geraniol (29) and (E)-ocimene (30) are largely responsible for the high degree of diffusion, the vigour and pleasant freshness of the scent (cf. analysis p. 199).

Brassavola tuberculata

is also an exclusively night-scented species. Otherwise known as *Brassavola fragrans,* it grows epiphytically in the coastal regions of Brazil and Bolivia.

Its 'white-floral' perfume is also highly diffusive, reminiscent of night-scented *Nicotiana* species and tiaré *(Gardenia taitensis),* and contains—in addition to fairly large amounts of ocimene—methyl benzoate, methyl salicylate and phenylethyl acetate as its main components (cf. analysis, p. 200). Two *Brassavola* species form a striking contrast with those just described, both from the morphological and olfactory standpoints, and this has led certain taxonomists to assign

50

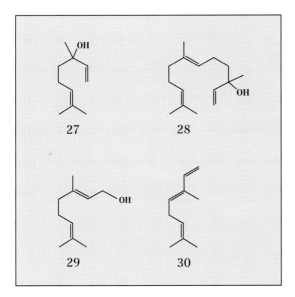

them to a separate, specially created genus—'Rhyncholaelia'. The species in question, discussed below, are B. glauca and the similarly scented and closely related B. digbyana.

Brassavola glauca

grows in the sparse, well-lit mountain forests of Mexico, Guatemala and Honduras at altitudes between 1200 and 1500 m. It grows epiphytically on the bare bark of oak trees, or lithophytically on stony ground. This rare species already differs from the others in the genus by its very vigorous growth. Its name derives from the bluish-green colouring of the leaves, which is produced by a fine waxy layer. But its most surprising feature is its scent, first appearing after dusk and increasing in intensity until about 23 hr. Its main components are (E)-ocimene, citronellol and nerolidol.

This scent could certainly not be characterized by the term 'white-floral', rather triggering clear associations with the 'rosy-floral' notes of lily of the valley and cyclamen *(Cyclamen purpurascens)*, and incorporating only very weak hints of 'white-floral' notes. As shown by the analysis on page 199, this scent, so untypical of a night-scented orchid, represents a combination of the basic skeletons of 'rosy-floral' and 'white-floral' scents, and the concentrations of the olfactorily important compounds are such that the 'rosy-floral' aspect predominates. Irrespective of the existing theories and experiences, these

51

olfactory observations and analytical results suggest that these two species might have developed from day-scented representatives of the subtribe Laeliinae, and the 'rosy-floral' part of the scent represents a relic, as it were, of the formerly day-scented form.

As already briefly mentioned, the closely related

Brassavola digbyana

possesses a very similar scent composition (cf. analysis, p. 199). Its native territories are

Figures
50 *Brassavola nodosa* (L.) Lindl.
51 *Brassavola tuberculata* Hook.

Orchids of the American tropics

52

53

the hot savannas from Mexico to Belize, where it grows at altitudes up to 800 m. Because of its large flowers, the fully fringed lip and its strong, pleasant scent, *B. digbyana* is very often used for hybridization with *Cattleya, Laelia* and other closely related genera.

Brassia R.Br.

In stark contrast to the night-scented species of the *Brassavola* genus just described are the twenty or so representatives of the *Brassia* genus which grow epiphytically in humid, sunlit forests of tropical America. The inflorescence develops from the base of the pseudobulbs, and it can vary in length from a few centimetres to a metre. Each inflorescence bears from 6 to 20 flowers, arranged in two rows. Since the sepals are usually longer than the petals, narrowing to form sepaline tails, the brown, yellow and green coloured flowers look rather like spiders. A particularly impressive representative of the 'spider orchids' is the rare

Brassia verucosa,

which occurs in humid, sunlit forests at altitudes of 600–1600 m from southern Mexico to Venezuela. Each inflorescence, growing up to 80 cm in length, often bears between 15 and 20 large flowers with sepals as long as 12 cm.

Its characteristic scent, evoking a fern-covered heath and the resinoid secretion of *Cistus* species, is based on the earthy-camphoraceous linalool character of *cis*-linalool oxide which, together with the labdanum-like and herbaceous aspects of phenyl-propionic aldehyde, phenylpropyl alcohol and p-methyl acetophenone, form the characteristic '*Brassia* scent' (cf. analysis p. 200). This unusual floral scent is encountered in various *Brassia* species.

Cattleya Lindl.

In 1823, an avid English plant collector, William Cattley, received a parcel of plants

Orchids of the American tropics

from Brazil wrapped in strange, thick stalks serving as 'live packaging material'. Out of curiosity, Cattley planted some of these stems and placed them in a greenhouse. The stems eventually formed shoots and, much to his surprise, in the spring of 1824, one of the plants produced some splendid lilac-violet flowers. Being unable to identify this plant himself, however, he passed it on to his friend John Lindley, a botany professor and the then director of Kew Gardens. Lindley soon realized that he was looking at a new orchid genus and, in his description, in 'Collectanea Botanica', dedicated it to his friend Cattley by calling it *Cattleya labiata*.

Most of the orchids collected in Brazil at this time and sent back to England eventually passed through the hands of this famous botanist for the purposes of description and classification. On a visit to the 'Orchid Herbarium' at Kew Gardens, the sharp-eyed observer can still see, next to the electronic computerized systems, Lindley's original filing cabinets. Even in Lindley's time, these large lavender-pink-violet flowers attracted much attention, and the *Cattleya* genus remains, to this day, the best-known branch of the orchid family. In fact, these are often considered to be the quintessential orchids.

The 60 or so species of the *Cattleya* genus are distributed over a region extending from Mexico, across the whole of Central America and large parts of South America down to Uruguay, with the highest concentration in Brazil. Taxonomists are still unclear about the precise relationships between *Cattleya* and a number of other, closely related genera. Many species of the closely associated *Laelia* genus are so similar in their floral structure that they could easily be mistaken for *Cattleya* species. One important distinction feature is the number of pollinia: four in *Cattleya*, eight in *Laelia;* but once again this rule is not without its exceptions. Some of the *Cattleya* species produce only one leaf per pseudobulb—the so-called unifoliate Cattleyas; others produce two leaves—the bifoliate Cattleyas. This rough classification, often employed in the botanical literature, is also used in the following descriptions of the individual species.

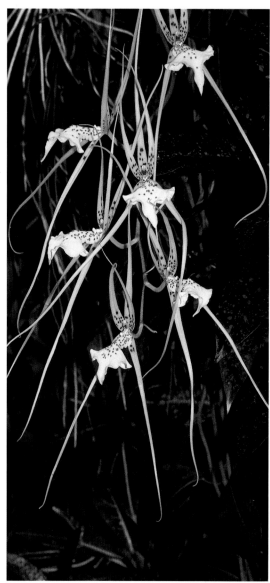

54

The unifoliate Cattleyas

This *Cattleya* group, also commonly known as the 'labiata group', has attracted much attention since the genus was first described by Lindley, though this would not surprise those who are familiar with these extraordinary flowers. No other orchid has been used as much for interbreeding with related genera,

Figures
52 *Brassavola glauca* Lindl.
53 *Brassavola digbyana* Lindl.
54 *Brassia verucosa* Lindl.

55

with the result that thousands of hybrids are now in existence. Their description is not, however, within the scope of this book.

In view of its history and magnificent appearance, it is not surprising that

Cattleya labiata

has been chosen as the type species of this genus.

These flowers, varying in the intensity of the lilac-pink shading of their sepals and petals and the magenta colouring and yellow markings of their lip, can grow to 17 cm in diameter when fully developed, and have come to symbolize the splendour and abundance of the Tropics. In perfect harmony with its visual appearance is its 'aromatic-spicy-floral' scent, whose composition is described in greater detail on page 203. This scent is also present, in various modified forms, in many related unifoliate Cattleyas. According to tradition, the first plant of this species to reach Europe originated from the Serra dos Orgaos north of Rio de Janeiro, but it has never been seen there since. However, one of its typical haunts is the mountainous region of Pernambuco, where a three-month rainy season is followed by a very hot dry season.

A characteristic feature of this 'Queen' of orchids is that the number of varieties is so great that it is almost impossible to find two identical plants, unless they have been cultured from the same clone by division. In genetic terms, *C. labiata* does not appear to be very stable, and over 70 varieties have been described to date. This state of affairs is also reflected in its scent which, though unmistakable from variety to variety, can vary considerably with regard to the intensity of different aspects. This applies in particular to the velvety woody note produced by caryophyllene epoxide, the hint of eugenol, reminiscent of cloves, and the floral-animalic note of indole, which in this case produces an almost 'crystalline' effect. As with most of the species in the 'labiata group', the flowers of all the varieties of *Cattleya labiata* exhibit differing shades of pinkish-violet. These flowers are all the more spectacular, therefore, when the typical colour is replaced by yellow and red, as is the case with the famous and highly-prized

Cattleya dowiana

of Costa Rica.

Often described as the most beautiful of orchids, *C. dowiana* provides a surprising contrast to the other species in the labiata group, not just because of its starkly contrasting colours but also through its distinctive and highly delicate scent. Over a heavy 'aromatic spicy-

Figures
55 *Cattleya labiata* Lindl.
56 *Cattleya dowiana* Batem. et Rchb.f.
57 *Cattleya lawrenceana* Rchb.f.

Orchids of the American tropics

58

floral' basic accord, attributable primarily to phenylethyl alcohol, 2-amino benzaldehyde, eugenol, indole and a touch of vanilline, hovers a very fresh, hesperidic and 'rosy-floral' top note, produced mainly by linalool, neral and geranial (cf. analysis, p. 203).

Slightly smaller, but no less eyecatching, are the flowers of

Cattleya lawrenceana,

an epiphytic orchid that grows from Guyana to Venezuela. Although its scent is usually rather weak, clones are occasionally found to emit a strong, velvety 'aromatic rosy-floral' scent, produced primarily by geraniol, ocimene and various aromatic esters, and accompanied, as in the case of *C. labiata,* by eugenol and indole (cf. analysis, p. 204).

Although only described by Lindley in 1831,

Cattleya maxima

is probably one of the longest known species of this genus. It was discovered in 1777 by the Spanish botanists Ruiz and Pavon in the Peruvian Andes. This species also grows epiphytically in Colombia and Ecuador in widely differing biotopes at altitudes between 500 and 1400 m. As with many representatives of the 'labiata group' it is subject to considerable variation, most strikingly in the intensity of the floral colouring, and it has also been described as weakly-scented, or even scentless. The illustrated examples, however, exude an extremely pleasant and appealing scent, giving off a hint of sweet pea and the 'labiata basic accord' at about 10 o'clock in the morning, and then greatly increasing in intensity and radiant strength until about 2 o'clock in the afternoon. Significant constituents of this scent include (E)-ocimene, methyl benzoate, methyl anthranilate and farnesol (cf. analysis, p. 205).

Although, like *C. maxima,* it belongs to the 'labiata group', and although it possesses extremely attractive flowers both as regards shape and colouring, one particular *Cattleya* produces a strikingly different scent. The species in question is

Cattleya percivaliana,

a mostly lithophytic orchid that grows on steep, sunny crags in the Venezuelan Andes at altitudes between 1400 and 2000 m.

Venezuelans would certainly have preferred this *Cattleya* to have been dedicated to Simon de Bolivar, the 'Great Liberator', rather than the English orchid enthusiast R.P. Percival. According to legend, whilst on a recuperative stopover in the Venezuelan Cordilleras, Bolivar was presented, by a poor woman, with a large flowering specimen of

Figures
58 *Cattleya maxima* Lindl.
59 *Cattleya percivaliana* O'Brien

Orchids of the American tropics

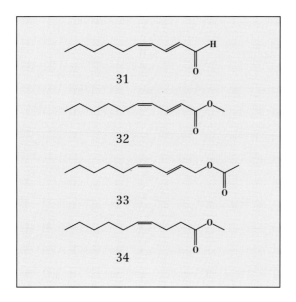

C. percivaliana. He was so filled with admiration for the plant that he requested that it be used as a symbol of the beauty of Venezuela, hence its other name of 'Flor del Libertador'. However, this splendidly coloured *Cattleya* with its deep orange throat was not a particularly popular choice. The ever familiar 'labiata basic scent' is accompanied by the dissonant accord that we encountered in the earlier section on *Chondrorhyncha lendyana* (cf. p. 52) and that was reminiscent of the defensive secretions of bugs and cumin seeds, with an additional very specific green-fruity facet. A range of double-unsaturated C_{10}-lipid metabolites is responsible for this highly characteristic scent aspect, including (E,Z)-2,4-decadienal (31), methyl (E,Z)-2,4-decadienoate (32), (E,Z)-2,4-decadienyl acetate (33), the corresponding (E,E)-isomers and single-unsaturated derivatives, such as methyl (Z)-4-decenoate (34) (cf. analysis, p. 205).

Compounds 31–34 are encountered fairly frequently in orchid scents, generally at rather low concentrations. This particularly applies to 2,4-decadienal (31), which has been identified in 11 out of 250 species. Methyl (E,Z)-2,4-decadienoate (32) also represents an important constituent of the aggregation pheromone of the bark beetle, *Pityogenes chalcographus* [26], and impressively illustrates how nature uses the same compounds for highly diverse purposes.

Two species of the unifoliate Cattleyas are so unlike the rest of the group in appearance and scent that they are not included in alphabetical order, but are rather given as concluding examples. The first species demonstrates how even the most inaccessible regions of tropical primeval forest can harbor some exceptional plant and animal species of which we are totally ignorant.

Cattleya araguaiensis

was discovered in the 1960s growing along the Araguaia river in the depths of the Brazilian Amazon region. Though small in stature, its flowers are almost as large as the whole plant, reaching a diameter of about 10 cm.

The green, brown and red hued flowers emit a strange and rather subdued scent characterized on the one hand by aldehydic-green notes reminiscent of octanal, nonanal and farnesal and, on the other, by aromatic, spicy-floral, almost dusty, aspects. The latter aspect is mainly due to the combination of methyl benzoate, methyl salicylate, methyl cinnamate, eugenol and methyl anthranilate (cf. analysis, p. 202).

The second species differing dramatically from the rest of the 'labiata group' is the small, yellow-flowering and extraordinarily graceful orchid,

Cattleya luteola,

which grows in the Amazon basin and the adjacent regions of Peru and Ecuador. For years I had assumed *C. luteola* to be almost scentless until, in connexion with the investigation of a night-scented *Aerangis* species, I discovered that, during the early morning hours between 4 and 8 o'clock, it emitted a restrained, very delicate and transparent fresh-floral scent, complemented by leafy-green and woody notes. The intensity of the scent probably reaches a peak between 5 and 6 o'clock, declining sharply after 8 o'clock. This time-dependent scent release is particularly marked in flowers whose inner portion of the lip is coloured yellow, with no red markings. Flowers with the red markings emanate a certain amount of scent throughout the day, though they, too, show a peak in the early morning hours. As observed by Dodson during field studies in Peru [11], *C. luteola* is visited and pollinated by crepuscular bees between 5.30 and 5.45 in the morning. It would appear that scent release is extremely well synchronized with the pollinator's short visiting time.

Cattleya luteola not only serves as a very impressive illustration of the need for careful monitoring of the floral scents over a full 24-hour cycle, but also demonstrates that the intensity of the scent, as experienced by the nose, is in no way indicative of the quantity of scent. Depending on the clone, sesquiterpene hydrocarbons, with comparatively high threshold levels, represent 85–95% of the scent, with caryophyllene forming the predominant component. The compounds that actually determine the scent—in particular *cis*-3-hexenol, nonanal, decanal, phenylethyl alcohol, jasmone and caryophyllene epoxide—account for only a small percentage (cf. analysis, p. 205).

The bifoliate Cattleyas

Although the bifoliate Cattleyas cannot compete with the unifoliate Cattleyas as regards flower size, they are, nevertheless, just as varied and interesting in respect of colouring, floral shape and scent. Their distribution is similar to that of the unifoliate Cattleyas, with the heaviest concentration in Brazil. The first of the species to be addressed,

Cattleya bicolor,

was first discovered, and subsequently painted, by M. Descourtilz in the Brazilian province of Minas Gerais. In this single instance, Lindley based his description of a flowering plant, published in the 'Botanical Register' of 1836, not on personal study, but on Descourtilz's painting. Live specimens only reached Europe several years later. This species, endemic to Brazil, grows epiphytically on trees at altitudes of 1000–1300 m, particularly in cool, damp river gorges, and can vary considerably in size and intensity of colouring. The literature describes several subspecies and varieties, whose evaluation and classification can often perplex even the specialists. Their scents cover the whole gamut from scentless to strongly-scented. The specimen illustrated is *C. bicolor,* which has spots on the sepals and petals visible as slight shading and emits a strong, aromatic-floral scent, incorporating a refreshing hesperidic note. This is due mainly to methyl benzoate, methyl salicylate and methyl cinnamate in combination with neral, geranial and eucalyptol (cf. analysis, p. 203).

Figures
60 *Cattleya araguaiensis* Pabst
61 *Cattleya luteola* Lindl.

Another variety of *C. bicolor,* though very similar in habitat and floral colouring, does not possess spots on the sepals and petals, and its scent—though also very similar—is weaker by a factor of at least 10. Yet another type, also strongly-scented and with particularly intense colouring (it has a dark crimson lip) is often known as *C. bicolor* var. *brasiliensis.* Olfactory evaluation of the scent suggests that it contains considerably more vanilline, eugenol and related compounds than the specimen described above.

Like the unifoliate Cattleyas, the bifoliates exhibit not only morphological relationships, but also, to a certain extent, olfactory interrelationships. The aromatic-floral note of *C. bi-*

color, for example, is strongly reminiscent of the subsequently described *C. leopoldii,* although the fresh hesperidic note formed by neral, geranial and eucalyptol is lacking. Neral and geranial are present, however— and in fairly large quantities—in the scents of *C. granulosa* var. *schofieldiana* and *C. porphyroglossa,* both described below, whilst we have not yet been able to identify eucalyptol in any other bifoliate Cattleya.

Although

Cattleya granulosa

was described by Lindley in the 'Botanical Register' (1842) as originating from Guatemala, it is actually a native of the coastal region of the Brazilian federal state of Pernambuco, where it grows epiphytically on trees. The original mistake was probably due to the fact that the orchid shipment in question reached England via Guatemala and that the 'orchid hunters' of the time liked to conceal the original locations of their plants. The attractive flowers, reaching up to 11 cm in diameter, emanate a rather subdued, though moderately strong, scent, forming a pleasant accord of green fruity, 'rosy-floral', aromatic and balsamic aspects. This scent description also applies to the variety *schofieldiana,* illustrated here, though the floral, aromatic and balsamic aspects (produced by nerol and hydroquinone dimethyl ether) are even more in evidence. This variety is often used for crossings, and its distribution area, the federal state of Espíritu Santo, where it grows epiphytically between 400 and 700 m, differs from that of the nominate form.

Another native of Brazil,

Cattleya porphyroglossa,

belongs to the same group of orchids. As a result of the indiscriminate and excessive deforestation of its preferred biotope and the greed of collectors, this has become a highly endangered species. The original, relatively large distribution area of this rare plant is now virtually limited to the Paraíba estuary, where it grows epiphytically on trees, particularly in the area prone to flooding during the rainy season.

Photo Ch. Weymuth

The flowers—rather small for the *Cattleya* genus, reaching a maximum 7 cm in diameter—emit a very sweet 'rosy-floral' scent, but with a slight hint of the dissonance so characteristic of *C. percivaliana*. Of particular importance to these olfactory aspects are large quantities of nerol and geraniol, accompanied by smaller amounts of the two isomers of 2,4-decadienal (cf. analysis, p. 206).

One of the best-known bifoliate Cattleyas,

Cattleya leopoldii,

is very closely related to the somewhat smaller-flowered *Cattleya guttata,* and is often described as a variety of that species. *Cattleya leopoldii* grows in trees, at heights of up to 20 m, in the coastal region of southern Brazil between Paranagua and Espíritu Santo. The flowers, with their richly contrasting colours, grow to a maximum diameter of 11 cm and, at midday, emit a strong aromatic-floral scent characterized by a strangely sweet and heavy aspect. This is attributable to relatively large quantities of methyl anthranilate, indole and vanilline (cf. analysis, p. 204).

Figures
62 *Cattleya bicolor* Lindl.
63 *Cattleya granulosa* var. *schofieldiana* (Rchb.f.) Veitch.

Orchids of the American tropics

64   Photo Ch. Weymuth

The final bifoliate Cattleya to be described may already be extinct in the wild:

Cattleya schilleriana

is closely related to *C. leopoldii,* and also displays strong contrasts in its colouring. It was cultivated for the first time in 1857 by a Hamburg consul and orchid collector named Schiller, and the species was named after him by Reichenbach. The very small distribution area of this species is located in the Brazilian federal state of Espíritu Santo, in the district of Rio Jucu, where it grows (or grew) at altitudes of 170–800 m. Although the literature indicates that *C. schilleriana* is only weakly-scented, the clone investigated is one of the most strongly-scented orchids. Should the sniffer approach the flowers too closely, he will be accompanied for the next five minutes by their exceptionally heavy and sweet scent. This is largely due to the interaction of large quantities of geraniol with 2-amino benzaldehyde, indole and vanilline (cf. analysis, p. 206).

Caularthron Raf.

This genus, closely related to *Epidendrum,* encompasses about six species native to South America, Trinidad and other islands in the area. The thick vigorous pseudobulbs of these epiphytic orchids are often hollow and inhabited by ants, which enter through a small opening at the lower end of the pseudobulbs. By far the most renowned species is

Caularthron bicornutum,

whose native territory covers Colombia, Venezuela, Trinidad and Guyana (also known as the 'virgin orchid'). The long inflorescence emerges from between the shiny green leathery, lanceolate leaves, and bears 5 to 20 ivory-white flowers whose characteristically shaped lip is embellished with red spots.

The scent of this 'virgin orchid' awakens long-forgotten olfactory experiences from one's youth. Was it the first encounter with a

65   Photo Ch. Weymuth

66

perfume or the flavour of a sweet? Suddenly an image springs to mind, recalling all the sounds, smells and colours of the fair. Of course! Candy floss, with its strong and unusual aroma, evoking strawberries, raspberries and 'tutti frutti' all at the same time. Indeed, its scent analysis, described on page 206, could even provide the formula for a junior flavourist's 'fantasy fruit'. This particular scent is based on an 'ionone complex', composed primarily of β-ionone, geranylacetone and pseudoionone, together with linalool and aromatic compounds such as methyl salicylate, methyl cinnamate and cinnamic aldehyde.

Cirrhaea Lindl.

This genus includes approximately eight very similar epiphytic and lithophytic species which are all endemic to Brazil. Although, according to the descriptions in the literature, these bizarre plants can form large colonies in their natural habitat, they have nevertheless remained precious collectors' items. The in-

Figures
64 *Cattleya porphyroglossa* Linden et Rchb.f.
65 *Cattleya leopoldii* Lem.
66 *Cattleya schilleriana* Rchb.f.

Orchids of the American tropics

67

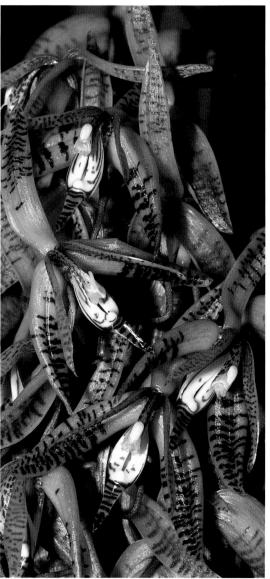

68

florescence grows from the base of the pseudobulbs and hangs downwards, bearing 5–20 tightly packed non-resupinated greenish-yellow to reddish-brown coloured flowers. The species most likely to be found in collections is

Cirrhaea dependens,

whose flowers give off a highly diffusive, fresh 'spicy-floral' and slightly herbaceous scent, composed of familiar orchid scent components such as myrcene, eucalyptol, a large quantity of (E)-ocimene and eugenol, all assembled in a particular ratio to constitute the '*Cirrhaea* character' (cf. analysis, p. 207).

Constantia

An even smaller, but no less fascinating, genus is *Constantia,* which includes four small, very rare, epiphytes, also endemic to Brazil. Their tiny pseudobulbs, measuring only 6–10 mm in diameter, grow so closely together as to form cushion-like colonies. From the apex of each of these bulbs emerges a pair of well-rounded, almost fluorescent-green leaves. In their natural habitat, they have a particular affinity for the trunks and branches of *Vellozia* species, and it seems to be impossible to cultivate them on other substrates. The most familiar representative is

69   Photo Ch. Weymuth

Constantia cipoensis,

from the Serra do Cipo mountains in Minas Gerais. This orchid is occasionally seen growing in collections on *Vellozia* cuttings. The very elegantly-shaped flowers are cream coloured with moss-green shading, and the diamond-shaped lip displays a contrasting yellowish-orange patch at its base.

This enchanting orchid is generally considered to be scentless, an attribute often all too glibly applied to many orchids. Closer acquaintance with this gem, however, is eventually rewarded with one of the most beautiful of orchid scents, which is in complete harmony with the shape and colouring of the flower. But this extraordinary mild, perfume-like scent can only be appreciated in the 40 or 50 minutes of twilight. Underlying the scent is a combination of 'ionone-floral' (β-ionone and derivatives), 'rosy-floral' (geraniol, geranylacetone, etc.) and aromatic-floral (benzyl acetate, methyl cinnamate, etc.) notes, rounded off by large amounts of linalool and considerable quantities of geranial (cf. analysis, p. 209).

While *Cattleya luteola* (cf. p. 67), which emits scent only during the dawn hours (when it is visited and pollinated by crepuscular bees), *Constantia cipoensis* also gives off its scent during a very limited period of the day and would appear to attract a bee species that is only active at dusk. Very similar behaviour is exhibited by *Masdevallia laucheana* (cf. p. 103), which emits a rather different, though equally fascinating scent.

Coryanthes and other typical 'euglossine orchids'

As briefly mentioned on p. 33, some of the strongest smelling orchids within certain neotropical genera are visited and pollinated exclusively by male euglossine bees. This fascinating scent-based relationship between orchid flower and euglossine bee has been investigated over the last 30 years by various researchers [11–14], and recently Gerlach and Schill [27] published a summary compilation of the scent compositions of many 'euglossine orchids'. To illustrate these exceptionally interesting orchids, several representatives of the highly typical genera *Coryanthes*, *Gongora*, *Stanhopea* and *Catasetum* will now be described.

Coryanthes Hook.

This genus, which has attracted considerable attention in recent years, includes 15–20

Figures
67 *Caularthron bicornutum* (Hook.) Raf.
68 *Cirrhaea dependens* (Lodd.) Rchb.f.
69 *Constantia cipoensis* C. Porto et Brade

large epiphytes, distributed from Guatemala to Peru and Brazil, invariably in extremely hot and humid regions at altitudes of 0–1200 m. About ten species grow in the main distribution area of the corresponding biotopes of Peru. As many orchid collectors will have discovered to their cost, *Coryanthes* species often grow in symbiosis with very aggressive tropical ant species which form tree nests, using the wickerwork of the plant roots as a framework. As further revealed by expedition reports, these nests gradually form 'ant gardens' which are then colonized by other plants, such as *Peperomia* and Gesneriaceae species, producing the distinctive symbiotic biotope. Consequently, many *Coryanthes* species exhibit extrafloral nectaries on new shoots, flower buds and other plant parts. These nectaries supply nectar to the ants which, for their part, keep the plants' enemies at bay.

The pendulous inflorescence of the *Coryanthes* species emerges laterally from the base of the pseudobulbs and bears two or three grotesque-looking flowers which are characterized by a very pronounced lip and huge lateral petals. The latter are rarely seen in their fully-extended triangular state, since they often curl up just 30 minutes after blooming. The highly complex structure of the lip consists of three parts, the upper spherical or cap-shaped hypochile (cf. p. 36, Figure 27, section A), the extended, and often furrowed, mesochile (middle section) and the bucket-shaped epichile, which has given the whole genus the popular name of 'bucket orchid'. Shortly before flowering, this 'bucket' is filled with a dilute sugar solution by two faucet-like organs at the base of the column. Bees of the genera *Euglossa, Eupulsia* or *Eulaema* [11] are attracted by the intense scent, which emanates mainly from the hypochile, and they proceed to collect the orchid perfume from this extremely exposed site. During this operation they often slip on the waxy surface, fall into the fluid-filled bucket and are then unable to fly away because of their wet wings. Their only escape route is through a narrow tunnel between the bucket and the top of the column, where the pollinia ultimately attach

70 Photo Ch. Weymuth

themselves to the pollinator (Figure 27, section C).

A recently published paper by Gerlach and Schill [28] attempted to clarify the taxonomic relationships within the *Coryanthes* genus on the basis of the scent composition. In this project, a significant proportion of the known species was analysed. In the following, however, I shall restrict myself to three representatives of the genus.

Coryanthes leucocorys

was described by Rolfe over 100 years ago, and is therefore one of the longest known representatives of this genus. It is an inhabitant of the Peruvian province Loreto, and is easily

Photo G. Gerlach

differentiated from all other species by its large helmet-like hypochile, which covers all of the mesochile, and even part of the bucket-shaped epichile. Its scent is extremely intense, up to 80% being accounted for by the widespread floral scent component methyl salicylate. The analysis (cf. page 210) also shows a characteristic pattern of even-numbered aliphatic alcohols and their acetates. It may be this pattern which enables the methyl salicylate to achieve the required selectivity in respect of the pollinator.

Coryanthes speciosa

is one of the very variable and widely distributed species within the genus, occupying a zone stretching from Guatemala to Brazil. The scent of the extremely striking flowers is usually strongly marked by eucalyptol. However, the scent of *Coryanthes picturata,* a species from Belize and Costa Rica now considered to be a 'variant' of *C. speciosa* deviates sharply from that of the normal form. It appears terpene-like, rather musty and somewhat leathery and dusty, but also has a very characteristic fresh aspect which, as only the 'initiated' would know, derives from the 1,3,5-undecatrienes.

The analysis on p. 210 confirms that the scent of this *Coryanthes* does, in fact, contain a small quantity of (E,Z)-1,3,5-undecatriene, in addition to the principal components α-pinene, β-pinene and sabinene. The charac-

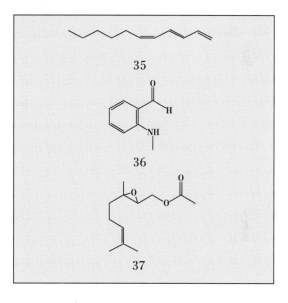

teristic leathery and dust-like scent aspect is due to the presence of 2-N-methylamino benzaldehyde in concentrations up to about 10%. This interesting derivative of 2-amino benzaldehyde, a substance which occurs widely in nature, was found by Gerlach and Schill [27] to be a dominant component in the scents of three varieties of *C. speciosa* that grow in Colombia (department Choco). However, this compound (36) also accounts for over 80% of the scent of *C. mastersiana* (cf. Figure 27,

Figures
70 *Coryanthes leucocorys* Rolfe
71 *Coryanthes picturata* Rchb.f.

Orchids of the American tropics

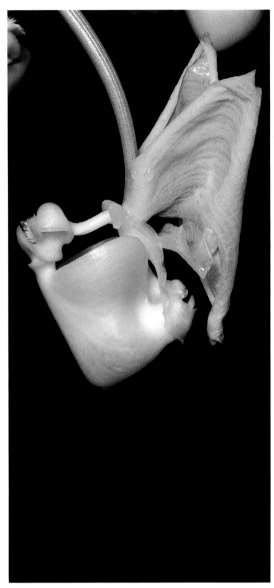

72   Photo G. Gerlach

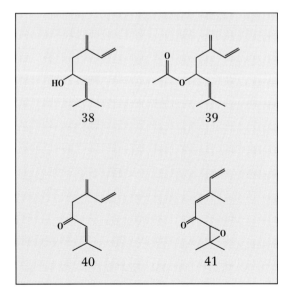

p. 36) as well as that of other *Coryanthes* species not yet identified, and occurs in lower concentrations in *Polystachya cultriformis* (cf. p. 141).

Equally worthy of mention is the olfactorily less significant compound 2(3)-epoxygeranyl acetate (37). This new natural substance, together with its precursor geranyl acetate, can also be detected in the floral scents of *Huntleya heteroclita* (p. 56), *Gongora truncata* (0.5%) and *Cattleya granulosa* var. *schofieldiana* (p. 68).

The third representative species to be mentioned also illustrates very clearly the great diversities of this genus with regard both to overall appearance and scent. I refer to the very rare, white-flowering

Coryanthes vieirae,

a native of Colombia, and just recently described by Gerlach [49]. Its scent is reminiscent of the floral note of linalool, accompanied by slightly woody notes. This verbal description corresponds largely with that of ipsdienol, which accounts for up to 77% of the scent. This trienol, 38, derived from myrcene and first identified by Silverstein et al. [29], is known to be a biologically active compound in the aggregation pheromone of many bark beetle species. In the scent of *Coryanthes vieirae*, ipsdienol (38) is accompanied by the corresponding acetate (39), ipsdienone (40) and (E)-2(3)-epoxy-2,6-dimethyl-5,7-octadien-4-one (41), a newly identified natural substance (cf. analysis, p. 210).

As recently reported by Whitten, Hills and Williams [30], ipsdienol occurs very widely in scents of 'euglossine orchids'. The derivatives 39–41 have yet to be described.

Gongora Ruiz et Pavon

The *Gongora* genus and the subsequently described genera *Stanhopea* and *Catasetum* possess exceedingly complex flowers that have fully adapted themselves to the respective euglossine bees. The basic principle of pollination with *Gongora* and *Stanhopea* is similar to that for the genus *Coryanthes*. The

76

bee alights on the lip, attempts to gain a secure foothold so that it can collect the all-pervasive scent, slips during the process and, whilst passing over the male part of the column, picks up the pollinia, which adhere to its abdomen. When this adventurous operation is repeated, the transported pollen is deposited in the stigma of the second flower.

Like the genera *Masdevallia* and *Anguloa*, discussed below, the genus *Gongora* was first mentioned as early as 1794 by the Spanish botanists Ruiz and Pavon in their 'Prodomus Florae Peruvianae et Chilensis', and they named it after Don Antonio Caballero y Gongora, the then Viceroy of New Grenada (Colombia and Ecuador). It includes about 25 epiphytic species, distributed from Mexico and the West Indies to Peru and Brazil. The inflorescence of the *Gongora* species always sprouts from the base of the pseudobulbs, is invariably pendulous and can bear up to 30 waxy, glossy, strongly-scented flowers. A species often seen in orchid collections is

Gongora armeniaca,

which is encountered in Nicaragua, Costa Rica and Panama at altitudes of 500–1500 m. The precise reason why Lindley should have chosen *armeniaca* as the species name for this plant when he first described it remains a mystery to this day. The inflorescence, reaching up to 40 cm in length, can bear up to 30 yellowish-brown to orange coloured flowers. These emanate an intense terpene-like scent, so typical of *Gongora* species of the *Acropera* section, which is only slightly floral, but is warm-herbaceous and almost labdanum-like. These latter aspects are produced by the combination of hydroquinone dimethyl ether, methyl 3-phenylpropionate and methyl cinnamate in particular, whilst (E)-ocimene and certain sesquiterpenes—present as the main components—contribute the terpene-like, slightly woody and balsamic notes (cf. analysis, p. 224).

Closely related in morphological terms, and possessing a very similar scent is

Gongora cassidea,

although in this case the scent is fuller and

73

more diffusive, with a marked floral aspect due to the much higher content of linalool (cf. analysis, p. 224).

The scent of *G. cassidea* is also characterized by a number of additional compounds. Particularly striking is the presence of 2-methylbutyraldoxime (2), plus derivatives, and phenylacetaldoxime (4) with its olfactorily important derivatives phenylacetaldoxime-O-methyl ether (42) and 1-nitro-

Figures
72 *Coryanthes vieirae* Gerlach
73 *Gongora armeniaca* Lindl.

2-phenylethane (43). As already mentioned on page 30, such degradation products of amino acids, including their derivatives, are very typical of the scents of night-scented flowering plants. In this respect, *G. cassidea* and the subsequently discussed *Bulbophyllum lobbii* (p. 146) are two notable exceptions. Equally worthy of mention is the presence of (E)-farnesene epoxide (44) which, after linalool, is the largest single component in the scent of *G. cassidea* with a content of 13.5%, and which is accompanied by the corresponding (Z) isomer in the ratio of 40:1.

We have also been able to identify this new natural substance in the scents of *Gongora armeniaca* (1.4%), *Gongora galeata* (1.7%) and *Coryanthes vieirae* (3.8%) (p. 76).

Stanhopea Hook.

The *Stanhopea* genus occupies a wide-ranging habitat extending from Mexico to Peru and Brazil, and, strictly speaking, incorporates about 25 species including some of the most impressive orchid species of the New World, if not of the whole orchid family. The inflorescence sprouts from the side of the relatively small, single-leaved pseudobulb. The flowers, which are often huge, invariably grow straight down. Their distinguishing features are a narrow column, which also points downward, and the opposing lip, which looks as if it has been turned on a lathe. Here, too, the lip has an extremely complex structure (cf. Figure 76). Instead of evolving into the helmet-like structure of *Coryanthes,* the throat of the lip, or hypochile, consists rather of a pouch-like depression. The central portion of the lip, the mesochile, presents two horn-shaped projections, whilst the terminal section of the lip, the epichile, forms a kind of funnel opening together with the tip of the column. Once again, every precaution is taken to ensure that the collecting operations of the euglossine bee around the hypochile are suddenly terminated by the bee sliding down through the funnel opening and brushing against the pollinia and stigma.

74

But the grandiose splendour of the *Stanhopea* flower is short-lived, lasting just one or, at most, three days, although it does emanate an overwhelming quantity of scent during this period. If the previously described trapping technique (page 22) is employed, trapping times of a few minutes often suffice to obtain high quality analytical samples.

As with many other neotropical genera, the flowers of the *Stanhopea* species can vary considerably in both colouring and morphological aspects. This explains why some 500 species names have been published to date for *Stanhopea*, although the taxonomists have managed to reduce the number of species to 25.

The first representative of this extremely fascinating genus,

Stanhopea ecornuta,

is a native of Guatemala, Honduras, Nicaragua and Costa Rica.

Its large, attractively-coloured flowers emit a highly diffusive scent, representing a perfect combination of fresh-floral and spicy-cinnamic notes, and largely attributable to the dominating main components (E)-ocimene and cinnamic aldehyde.

The second representative of this genus, the very elegant and, again, strongly-scented

Stanhopea jenischiana,

also possesses a definite cinnamic note, accompanied in this instance by fruity-balsamic aspects that could only originate from methyl cinnamate. In fact, this scent contains over 93% methyl (E)-cinnamate, accompanied by approx. 1% of the corresponding (Z)-isomer (cf. analysis, p. 244).

As pointed out by Williams and Whitten [31], the scent of *Stanhopea embreei* possesses almost the same composition. *Stanhopea jenischiana* is a native of Peru,

75

76

Figures
74 *Gongora cassidea* Rchb.f.
75 *Stanhopea ecornuta* Lem.
76 *Stanhopea jenischiana* Kramer et Rchb.f.

77

Colombia and Ecuador, where it generally grows epiphytically.

The scent of the third representative of this genus,

Stanhopea oculata,

is also completely dominated by a single component. In this case, the substance is eucalyptol, which accounts for over 85% of the headspace sample, and, naturally, is easily identified by the nose alone (cf. analysis, p. 244). This species can vary greatly in its flower colouring and is distributed from Mexico to Honduras.

The floral scents of the *Stanhopea* species investigated thus far almost invariably show a single, predominant main component, accompanied by a series of compounds contained in small quantities.

The same applies unreservedly to

Stanhopea tigrina,

which grows on the Atlantic side of Mexico at altitudes of about 2000 m as an epiphyte in humid forests. Its diffusive, 'aromatic spicy-floral' scent is extremely sweet. One would not expect it to contain more than 90% phenylethyl acetate; in fact, it is difficult even to detect this well-known substance in the scent of *Stanhopea tigrina*. The unusual accord is attributable to minor constituents accompanying phenylethyl acetate, in particular β-

78

ionone and derivatives, coumarine (46), p-hydroxy phenylbutanone (45) and vanilline (20) (cf. analysis, p. 244).

Compound 45, also known as raspberry ketone, together with vanilline (20) and coumarine (46), forms an extremely heavy and sweet aromatic-spicy accord that is probably unique among orchid scents.

Embreea rodigasiana –
formerly *Stanhopea rodigasiana*

Plate No. 7702 in Vol. 126 of the 'Botanical Magazine' (1890) showed a painting of this plant, which had been described two years previously by Cogniaux. R.A. Rolfe, the then curator of the Botanical Gardens at Kew, suggested that while it had the appearance of a *Stanhopea,* closer inspection revealed it to have certain morphological aspects not encountered in any other species of that genus. However, it was only some 80 years later that Dodson, at the suggestion of various orchidologists, finally transferred this orchid from the *Stanhopea* genus to the new genus of *Embreea,* which was specially created for this one species.

This orchid, rarely seen in collections, grows in Colombia and southern Ecuador at altitudes of 500–1000 m. Its scent, which is comparable in intensity with that of *Stanhopea* species, is eucalyptus-like, remi-

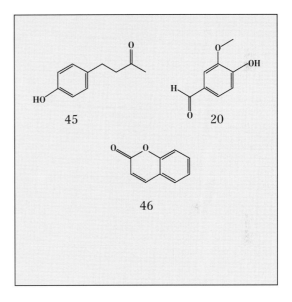

niscent of the aldehydic aspect of citrus fruits, and also possessing a warm, aromatic aniseed note. These impressions closely correlate with the main components eucalyptol, tetradecanal and hydroquinone dimethyl ether (cf. analysis, p.220).

Also worthy of special mention is the structurally and olfactorily interesting (Z,Z,Z)-3,6,9-dodecatrienol, which we first identified in *Jasminum sambac* [32] and which contributes its orange-like note to the scent of *Embreea rodigasiana.*

Figures
77 *Stanhopea oculata* (Lodd.) Lindl.
78 *Stanhopea tigrina* Batem.

Catasetum Kunth

The highly interesting genus *Catasetum* may be differentiated from the previously mentioned genera of 'euglossine orchids' by their unisexual flowers, which can reside on one and the same plant and yet flower at different times, thus rendering self-pollination virtually impossible.

A limited number of species can, in addition to the male and female flowers, also develop bisexual flowers. As a result of the failure to recognize these relationships, various forms of the same plant were originally even allotted different genus names. Darwin and Rolfe eventually clarified matters in the 'Journal of the Linnaean Society of London' (1862, 1891).

Interestingly, scent emanation in most *Catasetum* species only begins two to three days after flowering. The euglossine bees attracted by the scent then try to reach the scent, which forms particularly high concentrations below the wishbone antennae (cf. Figure 80) at the base of the column. As they make con-

Figures
79 *Embreea rodigasiana* (Claes ex Cogniaux) Dodson
80 *Catasetum pileatum* Rchb.f. (male flower)
81 *Catasetum viridiflavum* Hk. (male flower)

tact with these antennae, they trigger a mechanism that hurls the pollinia with great force onto the pollinator.

Depending on the reference source consulted, the genus *Catasetum* spans some 50–100 species covering the whole of tropical Central and South America, including the West Indies.

The first representative dealt with here,

Catasetum pileatum,

is known as the national flower of Venezuela, although it also grows in Brazil and Trinidad. This large-flowered epiphyte forms tightly-packed clusters of spindle-shaped or oval pseudobulbs bearing several leaves. The pendulous male inflorescence can grow up to 40 cm in length, and is densely covered with large flowers which measure up to 9 cm in diameter, and emit a caraway scent that is characteristic of many *Catasetum* species. This is formed by carvone, the typical caraway constituent, plus certain derivatives, particularly *trans*-carvone epoxide (47) (cf. analysis, p. 201). The latter substance was only recently described by Whitten, Williams and co-authors [33–34] as a scent component of this species, and of numerous other *Catasetum* species.

In the second species to be mentioned,

Catasetum viridiflavum,

trans-carvone epoxide (47) is also the domi-

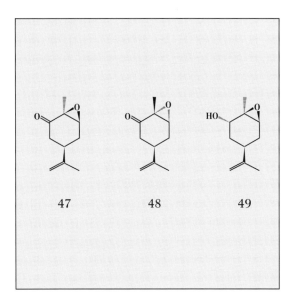

nant component in quantitative terms, with a content of 50%. In this case, it is accompanied by 0.2% of the corresponding *cis*-isomer 48 and 0.5% of a derivative of 47, *trans*-(*trans*-carveol) epoxide (49) (cf. analysis, p. 201).

The scent of *Catasetum viridiflavum* also contains fairly large quantities of thymol and indole, which modify the basic caraway accord to produce a very powerful spicy bouquet containing definite leathery aspects.

Dichea Lindl.

The genus *Dichea* includes 20–40 species that are all native to the hot, humid regions

Photo Ch. Weymuth

Orchids of the American tropics

of tropical America. These plants, seldom seen in collections, do not develop any pseudobulbs, but rather sprout shoots of widely differing length that are tightly packed with overlapping leaf bases. The generally small flowers grow individually along the leaf axis, and commonly emit a characteristic scent that is reminiscent of yeast. This particular note is also very apparent in the scent of *Dichea picta,* for example, and, to a slightly lesser extent, in that of the Brazilian orchid

Dichea rodriguesii.

As with *Pescatorea cerina* (p. 57) and *Cattleya percivaliana* (p. 65), this particular note is attributable to unsaturated C_{10}-lipid metabolites. Although these compounds, taken together, account for only 0.7% of the composition, they nevertheless make a significant contribution to the basic skeleton, which largely consists of benzyl acetate, phenylethyl acetate and, in the *para* position, the corresponding methoxylated derivatives (cf. analysis, p. 219).

Encyclia Hook.

The genus *Encyclia* encompasses some 150 species that are found predominantly in Mexico and the West Indies. This genus is very closely related to *Cattleya* (p. 60) and *Epidendrum* (p.88). In fact, it would be very difficult to distinguish it from *Cattleya* on the basis of morphological criteria alone. Rather, it is the extreme differences in flower size and the Mexican distribution that have led to this separate classification. *Encyclia* differs from *Epidendrum* in that it possesses pseudobulbs and a lip that is only partially, if at all, fused with the column. The floral structure of *Encyclia* species is also targeted mainly on species of small bees and wasps, whilst the genus *Epidendrum* shows much greater variations in its pollination mechanisms.

From the olfactory standpoint, the separate classification is definitely justified, particularly when compared with other genus allocations. One could even go so far as to talk of an '*Encyclia* scent', which could be characterized as a very diffusive 'ionone-floral' scent accompanied by distinctive 'aromatic-floral' and 'spicy-floral' notes. The sequence of these basic aspects may, however, be reversed. And of course there are exceptions, for example *Encyclia citrina* (p.86).

The first representative to be described here,

Encyclia adenocarpa,

is an epiphytic native of Guatemala that gives off a particularly ionone-rich scent, contain-

Figures
82 *Dichea rodriguesii* Pabst
83 *Encyclia adenocarpa* (La Llave et Lex.) Schltr.

Photo Ch. Weymuth

Orchids of the American tropics

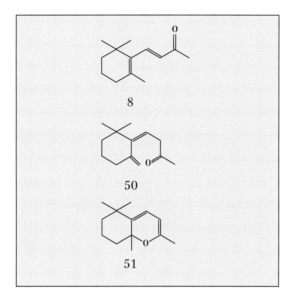

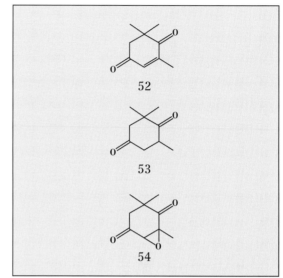

ing around 10% β-ionone (8), accompanied by its photocylization product (51), the so-called cyclic β-ionone, and the corresponding intermediate (Z)-retro-γ-ionone (50) [35] and other ionone compounds.

This high ionone content, together with the surprisingly high content of aliphatic and aromatic esters, form a scent accord that makes the presence of the extremely attractive orchid felt over a range of several metres (cf. analysis, p. 220).

The second species of this genus to be presented,

Encyclia baculus,

is actually better known by its other name *Encyclia pentotis,* and is found in almost every collection. It is distributed from Mexico to Colombia at altitudes of 400–1700 m. *Encyclia baculus* is characterized particularly by the completely symmetrical arrangement of its flowers, which are large for this genus and which also possess a strong scent. The relatively short inflorescence emerges from between the two leaves at the top of the pseudobulb.

Its scent is dominated by an 'aromatic spicy-floral' accord, consisting of aromatic esters, phenols, vanilline and indole, complemented by a distinctive herbaceous and straw-like note. This latter aspect is largely attributable to the large amount (27%) of oxoisophorone (52), accompanied by its dihydro derivative

84

53 and the corresponding epoxy diketone 54 (cf. analysis, p. 220).

Diketone 52 has been identified in 7 of the 250 orchid scents investigated, the scent of *Eria ovata* heading the list with a 55% content, followed by that of *Encyclia* baculus with 27%.

Encyclia citrina,

an orchid growing in the mountainous regions of Mexico, occupies a very special position in respect of both the appearance and scent, and is unlike any other species of this genus. It is restricted to altitudes between 2300 and 2600 m, particularly in the districts of Veracruz and Oaxaca, and notably in areas

86

Orchids of the American tropics

85

characterized by long, dry periods with negligible precipitation, cool winds and mist, but with plenty of light. *Encyclia citrina* was mentioned as early as the 17th century, under the Indian name of *corticoatz ontecoxochitl,* by a Jesuit priest named Hernandez who wrote a book on the natural history of Mexican animals and plants. He described how the plant was held in high regard by the natives.

As a defence against extreme climatic influences, the invariably downward-growing, oval pseudobulb, and the two leaves that sprout from it are coated with a whitish-grey wax layer. The very large— *for Encyclia*— waxy flowers possess a uniform yellow colouring and emit a unique, very pleasant floral and hesperidic scent which, essentially, is as unusual as the whole plant itself. It is based on an accord produced by the interaction of large quantities of ipsdienol (38) and ipsdienone (40), together with neral and geranial. The scent is elegantly rounded off by numerous olfactorily important compounds such as myrcene, citronellal, methyl geranate, methyl (Z)-4-decenoate, geraniol and farnesal (cf. analysis, p.221).

The next species in this genus to be discussed,

Encyclia fragrans,

is fairly closely related to the previously mentioned species *Encyclia baculus*. It has a similar distribution area, stretching from Mexico across Central America and the West Indies to northern South America, and has an affinity for similar biotopes. The characteristic difference is that *Encyclia fragrans* only produces one leaf per pseudobulb.

As early as 1782, *Encyclia fragrans* was flowering in the collection at Kew Gardens, and was, therefore, one of the first epiphytic orchids to be cultivated successfully in European collections.

As its name suggests, it is characterized by a powerful and very pleasant scent, and its relationship with *Encyclia baculus* is, to a certain extent, reflected in its aromatic-floral aspect. But the scent of *Encyclia fragrans* is more elegant, more balanced and multifaceted. A top note reminiscent of passion fruit and mango harmonizes admirably with the aromatic-floral basic accord, whilst an attractive contrast is formed by a rather astringent note that is reminiscent of certain tea roses. This latter aspect is produced primarily by 3,5-dimethoxy toluene (55), a substance also contained, for example, in the scent of *Chondrorhyncha lendyana* (p. 221),

Figures
84 *Encyclia baculus* (Rchb.f.) Dressler et Pollard
85 *Encyclia citrina* (La Llave et Lex.) Dressler

86

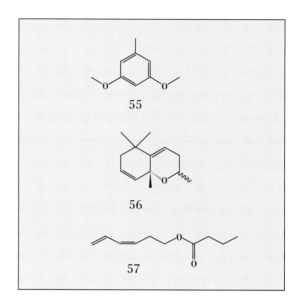

whilst the association with mango and passion fruit is triggered by the interaction of ocimene, β-ionone and the two isomers of edulane (56), plus a range of aliphatic esters (cf. analysis, p. 221).

Edulane (56), a substance that is relatively widespread in nature, and the highly typical ester (Z)-3,5-hexadienyl butyrate are, in fact, already known as important flavourings in passion fruit [36].

A few species of *Encyclia* are also found in South America. The best-known is probably

Encyclia glumacea,

a native orchid of Brazil, whose scent could not be mistaken for any other species. It is based on linalool, and its high anis aldehyde content produces a very sweet, aromatic-floral effect. This is contrasted by a melon-like green note that is principally attributable to (Z,Z)-3,6-nonadienol (analysis p. 222).

Epidendrum L.

Even though 150 or so *Encyclia* species have been transferred to a separate genus, *Epidendrum* still probably constitutes the largest genus in the western hemisphere, with about 1000 representatives. Most grow epiphytically, others lithophytically or even terrestrially, and they can be found from North Carolina to Argentina, although concentrated mainly in Brazil.

In terms of pollination, *Encyclia* is largely oriented towards bee and wasp species that are active during the day, and therefore includes many strongly-scented varieties. *Epidendrum,* on the other hand, specializes mainly in pollination by moths, butterflies

Figures
86 *Encyclia fragrans* (Swartz) Lemée
87 *Encyclia glumacea* (Lindl.) Pabst

Orchids of the American tropics

87

Orchids of the American tropics

and birds. Consequently, the particularly strongly-scented species are found mainly amongst the night-scented epidendra, whilst most of the species visited by butterflies usually emit a rather unassuming scent, and the 'humming-bird epidendra', almost invariably red or orange-red in colour, are even scentless.

Epidendrum ciliare

is a commonly-occurring species found in the Caribbean and Central America, and is one of the few tropical orchids described by Carolus Linnaeus himself. In contrast with many other orchids, it has adapted itself to widely differing living conditions, and often grows in large colonies on sun-drenched rock faces or, epiphytically, on a variety of tree species. The very slender flowers, distinguished by their white, three-lobed fringed lips, can grow to a diameter of 10 cm. Like some of the African moth orchids, they are almost scentless during the day but, as twilight approaches, they start emitting a very appealing 'white-floral' scent, accompanied during the first few hours by an astringent, grapefruit-like note. This is attributable principally to 2-methylbutyraldoxime and related compounds, which are characteristic of night-scented flowering plants (cf. analysis, p. 222).

Epidendrum ciliare not only varies in its overall appearance, but its scent can also emphasize different aspects, depending on the

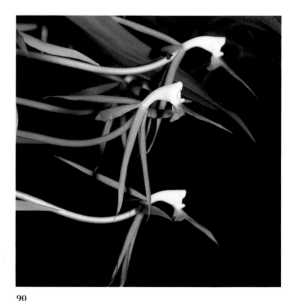

origin of the plant. It is precisely because orchids are so frequently subject to this variation that such great stress has been placed on the fact that the analyses described in Part three are based only on the scent of the illustrated plants.

As its name implies, and as might be deduced from the illustration,

Epidendrum nocturnum

also emanates its scent during the night. It is, however, quite out of the ordinary, since it shows no 'white-floral' aspects, but rather evokes eucalyptus and aniseed. This is due to the main component, eucalyptol, and its olfactorily important attendant substances, estragole, p-vinyl anisole and anethole (cf. analysis, p. 223).

This night-scented orchid, with its peculiar scent composition, was discovered around the mid-18th century on Martinique, and now forms a part of almost every orchid collection.

This very brief account of the *Epidendrum* genus ends with a particularly spectacular and rare species that is a native of the Chiapas district in Mexico and of Guatemala, namely,

Epidendrum lacertinum.

Although the overall impression created is similar to that of the two night-scented representatives of the same genus described above, this species reaches its peak scent emanation during the hottest hours of the day.

The scent is extremely difficult to describe. Apart from a green-floral note, it possesses an almost tingling shadow, producing a very transparent effect and changing into fresh air as the sniffer moves further away. The scent analysis described on page 223 partly explains why this scent is so difficult to define, being based on the combination of ocimene, lavandulyl acetate and indole.

91 Photo G. Gerlach

and partly as terrestrial orchids. The invariably large and robust plants develop single-leaved pseudobulbs and basal flower shoots up to 60 cm in length. These bear differing numbers of flowers, most of which are strongly-scented.

Houlletia Brongn.

The few, and very rare, species of the *Houlletia* genus are indigenous to tropical South America and grow partly as epiphytes

Figures
88 *Epidendrum ciliare* L.
89 *Epidendrum nocturnum* Jacquin
90 *Epidendrum lacertinum* Lindl.
91 *Houlletia odoratissima* Lind. ex Lindl. et Paxt.

One such species,

Houlletia odoratissima

fully lives up to its name.

A pronounced 'ionone-floral' scent, based on β-ionone, emanates from these chestnut coloured flowers with their bizarrely-shaped lips (cf. analysis, p. 226). In quantitative terms, however, the dominant compound is 7,11-epoxy-megastigma-5(6)-en-9-one (58), a substance that also occurs in large quantities in an unspecified *Gomesa* and *Gongora* species.

The epoxy-ketone 58 was identified by Näf and co-authors [37] as a minor component of passion fruit flavour and, in this context, has also been synthesized.

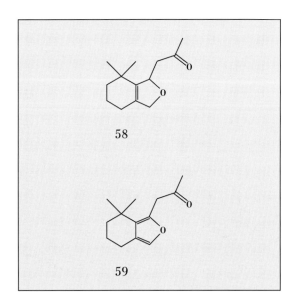

In the gas chromatogram of the scent of *Houlletia odoratissima*, a compound with a molecular weight of 206 is eluted shortly before 58. According to its mass spectrum, this is probably the corresponding dehydro compound 59, although its structure has yet to be confirmed by synthesis (cf. analysis, p. 226).

Laelia Lindl.

The genus *Laelia* was first mentioned by Lindley in 1831 in his famous work entitled 'Genera and Species of Orchidaceous Plants'. It is closely related to *Cattleya* and *Encyclia*, and many species in these three genera can easily be crossed. The 50 to 60 laelias that only occur in the Neotropics are distributed from Mexico to Brazil with, interestingly, two main areas of concentration—Mexico and Brazil. The very name *Laelia*, a girl's name in ancient Rome, anticipates the numerous attractive species in this genus.

The species

Laelia albida,

which grows on the Pacific side of Mexico, certainly lives up to such expectations. It tends to grow epiphytically on the bare bark of oak trees at relatively high altitudes (1600 to 2600 m), and occasionally co-exists with the no less attractive species *Encyclia citrina*. The normally white flowers, with an occasional hint of pink on top, emanate an aromatic 'white-floral' scent that is rather difficult to classify, although it also possesses 'rosy-floral' aspects and a top note reminiscent of melons. Of particular significance in this scent is the complex consisting of linalool, methyl benzoate, benzyl acetate, hydroquinone dimethyl ether and geranyl acetate. The hint of green-melon is provided by (Z,Z)-3,6-nonadienol and the corresponding acetate (cf. analysis, p. 227).

The following three lilac-coloured Mexican laelias—*Laelia anceps, Laelia autumnalis* and *Laelia gouldiana*—are closely interrelated, and it is occasionally suggested that *Laelia gouldiana* is the natural hybrid of the other two species.

Laelia anceps

grows in Mexico, on both the Gulf and Pacific sides, mainly at altitudes of 1000–2000 m in light oak forests interspersed with dry scrub. In these locations it can often be found growing alongside Bromeliaceae species and other orchids, e.g. *Brassavola glauca* (p. 59). *Laelia anceps* grows in large bushy clumps which, with as much as 50 inflorescences during the flowering period, can provide a real feast for the eyes. The species varies considerably in terms of size, flower colouring and lip markings, and at least 50 varieties have been described, although, in the opinion of the taxonomists, only a few of these can be classed as distinct varieties.

In olfactory terms, differences are particularly apparent between the normally lilac coloured types and the white-flowering 'alba forms'. The lilac coloured flowers have a rather unassuming, slightly terpenic, aldehydic and metallic floral scent, based largely on ocimene and characterized by trace components (cf. analysis, p. 227). The scent of the

Figures
92 *Laelia albida* Lindl.
93 *Laelia anceps* Lindl.
94 *Laelia autumnalis* Lindl.

Orchids of the American tropics

95

'alba forms' is usually somewhat stronger, fuller and more floral, and is reminiscent of *Laelia albida* and honeysuckle.

Possibly an even more familiar species of this genus is

Laelia autumnalis,

which inhabits more or less the same biotopes as *Laelia anceps,* occurring in the mountain regions of Oaxaca in the South, up to Sonora and Chihuahua in the North. This plant also forms large bushes and—both in its native territory and when cultivated in Europe—develops an inflorescence measuring up to 1 metre in length and bearing five to ten flowers of a vivid pinkish-lilac hue. Since they flower around All Saints' Day, these eyecatching blooms are known as 'Flor de Todos los Santos', and are sold in markets as bouquets at this time.

Their scent is considerably stronger than that of *Laelia anceps,* with a basic aromatic-floral accord and, in the top note, very delicate, green-aldehydic aspects, with an almost dissonant tone reminiscent of *Cattleya percivaliana* (p. 65). Indeed, the scent contains the two isomers of 2,4-decadienal and the corresponding acetates. But in this case they enhance the basic accord, which consists primarily of benzyl acetate, hydroquinone dimethyl ether, caryophyllene epoxide and cinnamic aldehyde, together with the green

note of (Z)-3-hexenol and (Z,Z)-3,6-decadienyl acetate (cf. analysis, p. 227).

Much theory and conjecture surrounds the third species of this group of Mexican laelias, the particularly splendid orchid

Laelia gouldiana.

For a long time it was assumed that all cultivated plants of this species were clones from a single natural plant that no longer existed in the wild. According to Withner [38] however, *Laelia gouldiana* can still be found as a rare plant growing in the cool and dry mountain regions of the state of Hidalgo.

In his book 'Orchidaceae of Mexico' (1951), L.O. Williams suggested that *Laelia gouldiana* was simply a more intensely-coloured variety of *Laelia autumnalis,* and therefore listed it as a synonym. On the other hand, Reichenbach's hypothesis that *L. gouldiana* might be a natural hybrid of *L. anceps* and *L. autumnalis* has recently gained increasing support.

A purely olfactory assessment of the scent of *L. gouldiana* does indeed evoke *L. autumnalis,* although the green, aldehydic note is less in evidence, while the overall scent appears to possess a greater concentration of 'ionone-floral' aspects. A comparison of the scent compositions of these three *Laelia* species does indeed show a very close relationship between *L. gouldiana* and *L. autumnalis* (cf. analysis, p. 228), whereas *L. anceps,* characterized particularly by ocimene, seems to stand on its own. Interestingly enough, ocimene could not be detected either in the scent of *L. gouldiana* or in that of *L. autumnalis.*

Laelia perinii

is certainly a worthy representative of those Brazilian species that form the second focal point of the genus. The first plant, still described by Lindley as *Cattleya perinii,* originated from the Serra dos Orgaos, but was subsequently found in the states of Minas Gerais and Espíritu Santo. The magnificent flowers can reach a diameter of 13 cm, and can vary considerably in their colouring.

The scent of the plant here illustrated is rather reminiscent of the typical 'labiata accord' of the unifoliate Cattleyas, with rather more emphasis on the mushroom-like aspects in the top note, accompanied by an additional, pleasant ylang note. These two olfactory impressions are principally attributable to 1-octen-3-ol and tiglyl benzoate (cf. analysis, p. 228).

Figures
95 *Laelia gouldiana* Rchb.f.
96 *Laelia perinii* Batem.

97

Lycaste Lindl.

The varying interplay of floral shapes and scents presented in this book is greatly enhanced by the extraordinary specimens in the *Lycaste* genus. Lindley was also captivated by the mysterious appeal of these flowers, naming the genus after Lycaste, the daughter of King Priam of Troy.

The principal distribution area of the 30 or so species, which are mainly epiphytic, but also partly lithophytic, is in the mountain ranges of Central America and the Andes. According to reports from collectors, those species not growing on trees seem to prefer the precipitous walls of ravines, growing right down to the watercourses below. The very thick pseudobulbs of these plants carry from one to three large, deciduous, plicated leaves. Several flower-bearing stalks sprout from the base, each of them usually bearing just one relatively large flower, typically characterized by large, splayed-out sepals and considerably smaller petals. The scents of the Lycastes cover a relatively broad range, the first representative,

Lycaste aromatica,

being one of the very strongly scented species of this genus. It is a native of rain forests from Mexico to Honduras, growing at altitudes up to 1200 m. The yellowish-orange flowers, often appearing in large numbers, possess a waxy texture, and emit a pleasant, spicy scent which is reminiscent of cinnamon and cloves but which also has refreshing floral aspects. This unmistakable scent is primarily attributable to the combination of methyl cinnamate (main component), cinnamic aldehyde, eugenol and ocimene (cf. analysis, p. 229).

Found at similar altitudes, and growing mainly on the Pacific side of Central America from Mexico to El Salvador,

Lycaste cruenta

is the second of the well-known yellow-flowering species of this genus. It can easily be differentiated from *L. aromatica* by its larger flowers, the characteristically-shaped lip and the less spicy but much more floral scent. Interestingly, the scents of these two species consist largely of the same individual scent constituents, although there are highly characteristic differences in the quantitative composition (cf. analysis, p. 229).

The compound responsible for the floral aspects, linalool, is present in much greater quantities in the scent of *L. cruenta,* whilst methyl (E)-cinnamate (60), a dominant component of *L. aromatica* in both olfactory and quantitative terms, occurs in minimal amounts in *L. cruenta*. In fact, the main components of the scent of *L. cruenta* are two derivatives of 60 methyl (Z)-4-methoxy-

Orchids of the American tropics

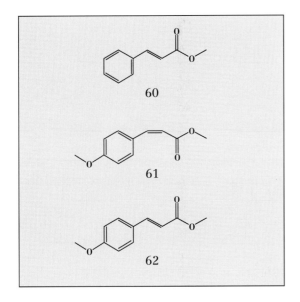

98   Photo Ch. Weymuth

cinnamate (61) and methyl (E)-4-methoxycinnamate (62). These methoxy cinnamates are employed in large quantities, in a modified form, as UV filters in cosmetic products, and possess only a very weak cinnamic note.

Perhaps one of the most fascinating species of this genus,

Lycaste locusta

is probably endemic to Peru. It generally grows terrestrially between rocks at altitudes of 2000–3000 m. Those fortunate enough to encounter this rare plant flowering in its native habitat or in a collection will understand why it bears the species name '*locusta*', from the Latin for grasshopper. In his original description in the 'Gardener's Chronicle' (11, 524, 1879), Reichenbach commented: "just as green as a green grasshopper or the dress of some Viennese ladies". Perhaps the colour of bottle-green or that of a Granny Smith apple would be even more appropriate.

Approaching these mysterious flowers, one is immediately struck by their scent. It is, in fact, very similar to that of a 'Granny Smith', and its composition, shown on page 230, does, indeed, reveal that this floral scent is almost indistinguishable from an apple aroma.

A complex of typical fruit esters, including ethyl acetate, ethyl 2-methylbutyrate, butyl butyrate and ethyl caproate is rounded off by smaller amounts of nonanal, decanal, α-ionone and β-ionone. We were very surprised to identify as the second largest component (confirmed by two control measurements) diethyl carbonate, whose biogenesis is rather difficult to explain.

Masdevallia Ruiz et Pavon

Following a virtual '*Masdevallia* mania' amongst collectors during the latter half of the 19th century—probably not unrelated to the rhapsodic original descriptions of Professor Reichenbach—very little was heard of these bulbless plants, which are made up of leaf clusters, until the 1960s. Since then, however, there has been a resurgence of renewed interest in this extremely varied genus, incorporating some 350 species and covering a distribution area from Mexico southwards through Central America and the whole of tropical South America. A particularly large number of species is found in the humid mountain forests of the Peruvian and Venezuelan Andes at altitudes above 2000 m. The renewed widespread enthusiasm for these plants is

Figures
97 *Lycaste aromatica* (Hook.) Lindl.
98 *Lycaste cruenta* (Lindl.) Lindl.

Orchids of the American tropics

Orchids of the American tropics

probably due largely to a major work by C.A. Luer entitled 'Thesaurus Masdevalliarium' (1983), which is greatly enhanced by the beautiful, artistic watercolours of Anne Marie Trechslin.

The sepals of the *Masdevallia* flower more or less fuse at the base to form a tube, and are invariably larger and more magnificently coloured than the petals and the lip. Indeed, in many species, the flower seems to consist only of three floral leaves, and since the tips of the sepals often taper to form tail-like structures, the overall appearance can be very bizarre. The first species of this genus was discovered as early as 1794 in the Peruvian Andes by Ruiz and Pavon, who named it *Masdevallia uniflora* in honour of their friend, the Spanish physician and botanist José Masdeval. Although this particular species is weakly scented, there is probably no other group of orchids that reflects the wide variety of orchid scents as well as *Masdevallia*.

As briefly mentioned on page 31, the genus *Masdevallia* incorporates a group of species which, with their putrid scent and their generally reddish-brown to yellowish-brown colouring, mimic rotting flesh, and thus attract carrion-feeding insects. A typical example is the previously mentioned species

Masdevallia caesia

(cf. Figure 22, p. 32), an orchid which is endemic to Colombia and which thrives as a pendulous plant in humid tropical mountain forests at altitudes of 2000–2500 m. Its unpleasant smell of butyric acid, rotting flesh and stinkhorn can in fact be ascribed, on the one hand, to butyric acid (63) and isovaleric acid (64) and, on the other, to large quantities of phenyl glyoxal (65), phenylethyl alcohol and α-hydroxy acetophenone (66) (cf. anal-

Figures
99 *Lycaste locusta* Rchb.f.
100 *Masdevallia elephanticeps* Rchb.f.
101 *Masdevallia striatella* Rchb.f.

Orchids of the American tropics

ysis, p. 230). It is difficult to believe that this scent, or rather stench, also contains vanilline.

Another example of this group is

Masdevallia elephanticeps,

which is also endemic to Colombia. Its stench is in no way less impressive than that of *M. caesia*. In this case, the particularly penetrating note is not produced by butyric acid, but by isovaleric acid, and although the sniffer might sense a hint of free amines, these have not been identified by instrumental analysis.

Rather more acceptable is the scent of

Masdevallia striatella,

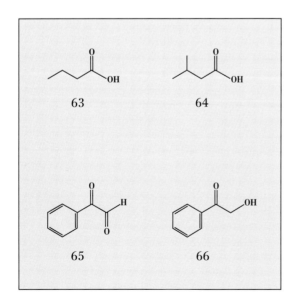

a species occurring in the tropical mountain forests of Venezuela at altitudes of 2000–2500 m. The relatively small flowers (1.0–1.5 cm in diameter) of this species, which grows epiphytically close to the ground, emanate a scent during the day that immediately triggers the association of the lion house at the zoo. After sunset, the animalic note is fairly weak, and a fresh-floral scent is more in evidence. In quantitative terms, this scent is based mainly on ocimene and linalool (cf. analysis, p. 231), a basic concept underlying many floral scents.

The pronounced animalic aspect of the scent of *M. striatella* is ultimately due to significant quantities of isovaleric acid and caproic acid. After sunset, however, the content of these two acids appears to decline markedly, with the result that the fresh-floral aspects come to the fore and the animalic note remains as a mere hint.

Also belonging to the group of rather unpleasantly-scented species is

Masdevallia tridens,

unmistakable in appearance and endemic to the tropical mountain forests of the western slopes of the Andes in Central Ecuador. The most striking morphological characteristic of *M. tridens* is the short, congested, horizontal raceme, with its comparatively large yellow flowers in mutual contact.

Orchids of the American tropics

shold and a rather weak scent (cf. analysis, p. 231).

The olfactory contrast with the floral scent of

Masdevallia estradae

could not be greater. This attractive Colombian species lives in the tropical mountain forest regions of New Grenada and Antioquia. The peculiarly-shaped flowers are dominated by the dorsal sepal that curves down over the widely spreading white, lateral sepals. At the same time, all three sepals form long, narrow sepaline tails, giving the flower, from a distance, an almost insect-like appearance.

Coming after the preceding species, its scent is an absolute delight. Within the orchid family, this scent can only be compared with that of the group headed by *Cymbidium goeringii (Cymbidium virescens)* (cf. p. 151). It is exceptionally pleasant, gentle and yet refreshing, evoking lily of the valley and, particularly, methyl jasmonate. In fact, I know of no other floral scent containing so much methyl *cis*-(Z)-jasmonate and accompanied so elegantly by farnesol, nerolidol, citronellol and linalool (cf. analysis, p. 230).

The *Masdevallia* genus also includes an outstanding example of a 'spicy-floral' scent. I refer to the absolutely transparent and, despite its warmth, almost crystalline scent of

Masdevallia glandulosa.

Its main component, eugenol, together with cinnamic alcohol and its derivatives, chavicol, vanilline, benzyl alcohol and phenylethyl alcohol, form the 'spicy-aromatic' basic accord which is complemented by the fresh-floral notes of linalool and ocimene and the fruity note of methyl 3-hydroxybutyrate.

This extremely eyecatching *Masdevallia* species, with its deep-purple flowers, was discovered just 13 years ago in northern Peru, growing at about 1200 m. It has managed to

'Its numerous, radiating sepaline tails call to mind the image of the sun's rays piercing through storm clouds'—a passage from C.A. Luer's description of these spectacular flowers. Their scent also conveyes a hint of this dramatic description, and is reminiscent of the dissonant aspect in the scent of *Cattleya percivaliana* (p. 65). In this case, however, the bug-like note stands out on its own since, apart from the compounds responsible, i.e. (E,Z)-2,4-decadienal (31), (E,Z)-2,4-decadienol and (E,Z)-2,4-decadienyl acetate (33), including the corresponding (E,E)-isomers, the only other compounds present—caryophyllene, for example—generally possess a high thre-

Figures
102 *Masdevallia tridens* Rchb.f.
103 *Masdevallia estradae* Rchb.f.

Orchids of the American tropics

105

104

find its way into many collections within this very short period.

Just as unmistakable, unique and as pleasant as the scents of *M. estradae* and *M. glandulosa,* is that of the last species to be presented,

Masdevallia laucheana.

The above statement could come as a surprise, since this species is considered by many orchid enthusiasts to be scentless. But should you chance to sniff these otherwise scentless flowers during the 30 or 40 minutes of twilight, you will be completely captivated by its extraordinarily diffusive, fresh 'rosy-floral' and 'ionone-floral' perfume. The main component, ocimene, is accompanied by neral, geranial, citronellol and geraniol and the resulting accord is rounded off by β-ionone.

As regards structure, the floral scent of this Costa Rican *Masdevallia* species contains two interesting new natural substances, although they are of secondary importance in olfactory terms. These are 3-oxo-7(E)-megastigmen-9-one (67), contained in relatively large quantities, and the corresponding hydroxy ketone (68) (cf. analysis, p. 231).

Satellites of the Masdevallia genus

It was Reichenbach himself who allocated a group of Masdevallias with saccular or shell-shaped lips to the so-called *Saccilabiatae* section. These were subsequently renamed Chimaeroideae by Kraenzlin in his monograph on the *Masdevallia* genus. He placed the particularly spectacular *M. chimaera* in the centre of this group, which was subsequently transferred wholesale by Luer in 1978 to the genus

Dracula Luer,

now comprising some 60 species. These orchids, most of which grow high up in the Colombian Andes, are unmatched by any other flowering plant in the extent to which they deviate from the common perception of a flower. The viewer is held spellbound in utter astonishment, shocked even, by the sight of these orchids. Little wonder, then, that the best-known species of this genus,

Dracula chimaera,

has been named after Homer's mythical beast with the head of a lion, the body of a goat and the tail of a serpent.

D. chimaera is endemic to the western Andes of Colombia, where it prefers dark, humid forests at altitudes of about 2000 m. Up to six

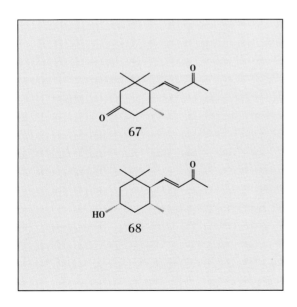

Figures
104 *Masdevallia laucheana* Kraenzl.
105 *Masdevallia glandulosa* Königer

Orchids of the American tropics

successive flowers can develop on each pendulous raceme. They give off a yeasty, algal, mushroomy scent that is typical of many *Dracula* species, accompanied, in *D. chimaera*, by certain cocoa-like aspects.

Many *Dracula* species are assumed to imitate the fruit bodies of pileated fungi, and thus attract fungus-fly species as pollinators [72]. This pollination principle is particularly apparent in another Colombian native,

Dracula chestertonii,

which emanates an unmistakably fungal aroma (cf. analysis, p. 219), and which has already been discussed in greater detail on page 31.

Another group of—in this case—short and small-flowered Masdevallias was also hived off at the same time by Luer to form the genus

Dryadella Luer.

These aptly-named orchids—one can easily imagine them as tree nymphs—differ from the Masdevallias in certain minor morphological aspects relating to the leaf and flower structure. As a typical representative of this group, I should like to mention

Dryadella edwallii

which, like many *Dryadella* species, is a native of Brazil.

Orchids of the American tropics

108

109

The scent of the numerous flowers nestling amongst the leaves is just like that of Iceland moss, and is especially due to the combination of limonene, 2-nonanone, anis aldehyde, creosol and acetic acid (cf. analysis, p. 219).

Maxillaria Ruiz et Pavon

Containing about 300 species, widely differing in both size and overall appearance, *Maxillaria* is one of the most extensive orchid genera in the neotropics. These plants grow epiphytically or lithophytically, and their distribution area stretches from Mexico right down to the southern part of South America. Some miniature species grow no longer than 4 cm, developing flowers measuring just a few millimetres in diameter. Other species form plants up to 1 m high, with flowers comparable in size with those of the unifoliate Cattleyas. As already mentioned in the section on scent and pollination principles (p. 29), by no means all orchid flowers pro-

Figures
106 *Dracula chimaera* (Rchb. f.) Luer
107 *Dryadella edwallii* (Cogn.) Luer
108 *Maxillaria nigrescens* Lindl.
109 *Maxillaria picta* Hook.

duce nectar. Moreover, since the accumulated pollen grains, the pollinia, cannot be utilized by the relevant pollinating insects, orchids have developed not only olfactory, but also various visual attractants. Thus, many orchids possess a powdery mass on the lip, so-called pseudopollen which, according to Pijl and Dodson [11], is, in fact, collected by certain bee species. According to the same authors, this phenomenon is demonstrated by at least 50 species of the *Maxillaria* genus.

In view of the size and variety of the genus, the following examples can only give a rough indication of their olfactory characteristics.

A very mysterious-looking species is

Maxillaria nigrescens,

which grows in poorly accessible precipices of the Colombian and Venezuelan Andes at altitudes of 2000–2600 m. The overall appearance of this large, epiphytic *Maxillaria* is so distinctive that it could not possibly be confused with any of the many other species in the same region. Several inflorescences, growing up to 15 cm in length, emerge from the base of the massive pseudobulbs, each of which bears one long, stiff, straplike leaf measuring up to 35 cm in length. Each inflorescence carries a large glossy purple-brownish-red flower. The special position of the sepals and the forward-leaning petals give it an unmistakable appearance.

Equally characteristic is its powerful, very diffusive scent, that can only adequately be described in analytical terms. Its main secret is the combination of β-ionone, including its derivatives, with (Z)-4-decenol and the extraordinarily intense (Z)-4-decenal (cf. analysis, p. 232). This complex is unique amongst the 250 orchid species investigated, and one would probably only need to experience it consciously on a single occasion in order to be able to recognize it again at any future date.

By contrast, a much more run-of-the-mill scent is the aromatic 'spicy-floral' perfume of

Maxillaria picta,

which is distinguished particularly by a note reminiscent of tarragon. It does, in fact, contain estragole as a main component, together with other compounds that contribute the floral and spicy aspects (cf. analysis, p. 232). The home of this species, commonly encountered in collections, is the Brazilian Serra dos Orgaos.

Also unmistakable are the overall appearance and scent of

Maxillaria tenuifolia,

which can be found growing from Mexico to Costa Rica in rain forests at altitudes up to 1500 m.

The flowers, with their striking, intense reddish-brown colouring, emit a highly typical

coconut scent, produced primarily by the main component δ-decalactone, accompanied by γ-decalactone and a series of methyl ketones (cf. analysis, p. 232).

The final species of the *Maxillaria* genus to be presented clearly serves to illustrate just how different the scent composition of colour variants can be. I refer to

Maxillaria variabilis,

a relatively common inhabitant of rain forests from Mexico to Panama. The flowers, just 7 mm in diameter, vary in colour from white to dark brownish-red. The two illustrated yellow and brownish-red types are particularly common. In order to obtain analysable scent samples from such small individual flowers, it is very important to miniaturize the trapping technique described on page 22. The scents of these two colour variants of *Maxillaria variabilis* are based on a common basic skeleton of (olfactorily rather insignificant) terpene hydrocarbons, consisting particularly of α-pinene, myrcene, limonene, α-copaene and germacrene D (cf. analysis, p. 233).

In the case of the yellow type, the characteristic scent is largely due to an aldehyde complex, based on the main components geranial and neral, and rounded off by citronellal, decanal and traces of (Z)-4-decenal. The very fresh-smelling complex is completely lacking in the scent of the brownish-red type, however, which is characterized rather by eucalyptol, 2-heptanol, aromatic esters such as methyl salicylate, γ-decalactone and small quantities of vanilline.

Miltonia Lindl.

The genus *Miltonia* encompasses about 20 epiphytic species that predominantly occur at the higher, cooler altitudes of the Andes of southeast Brazil. They are medium-sized plants with flat, compressed, yellowish-green pseudobulbs growing from a branching rhizome and each bearing two flexible light-green leaves measuring 20–50 cm in length. The inflorescences sprout from the base of the pseudobulbs and carry from one to ten medium-sized to large flowers.

Apart from various intra-generic hybrids, numerous inter-generic hybrids have also been developed, employing species from the genera *Oncidium, Odontoglossum, Brassia* and similar related genera. Such hybrids are commonly seen in flower shops. As far as

Figures
110 *Maxillaria tenuifolia* Lindl.
111 *Maxillaria variabilis* Lindl. yellow
112 *Maxillaria variabilis* Lindl. dark red

scent is concerned, however, it is the natural species that particularly deserve our attention. Take, for example, the long-flowering, very eyecatching

Miltonia regnellii,

which is encountered in the shady mountain forests of eastern Brazil. Just as their flowers could not be confused with those of any other species, the floral scent is also unique amongst the 2200 evaluated natural species. It is unmistakably reminiscent of the 'aldehydic' aspect of oranges and mandarins, and perhaps also of coriander. As the scent analysis on page 233 shows, this aspect is attributable to (E)-2-dodecenal, accounting for over 11% of the composition. This compound, commonly used as a flavouring is contained in small amounts in the peel of various citrus fruits and in coriander seed oil. The predominant component of this scent in quantitative terms is the structurally and biogenetically interesting (E)-4,8-dimethyl-1,3,7-nonatriene. This substance is contained in 30 of the 250 orchid scents investigated, and widely distributed, generally, in plant scents [16], though of secondary importance in olfactory terms.

By far the most famous *Miltonia*, on the basis of which Lindley established the genus in the 'Botanical Register' (1837), is the Brazilian native orchid

113

114

Miltonia spectabilis.

The variety illustrated, *moreliana,* is commonly seen in collections, and is distinguished by the intense colouring of its flowers.

The scent of these splendid flowers is particularly powerful around midday, giving a pronounced 'aromatic-floral' effect, and possessing a top note reminiscent of eucalyptus (cf. analysis, p. 234).

In this genus, *Miltonia warscewiczii* Rchb.f. and

Miltonia schroederiana

are exceptional in that they are distributed not in Brazil but in Central America. Since, moreover, they exhibit morphological differences compared with the *Miltonia* species just described, some botanists would tend to assign them to the genus *Odontoglossum,* discussed below.

The spectacular flowers of the rare *Miltonia schroederiana* emanate a characteristic, aromatic 'spicy-floral' scent, reminiscent of carnation and ylang-ylang, and based primarily on the combination of (E)-ocimene, cinnamyl acetate, cinnamic alcohol and eugenol (cf. analysis, p. 234).

Miltoniopsis Godefroy-Lebeuf

The genus *Miltoniopsis* was established as early as 1889 by Godefroy-Lebeuf in 'Orchidophile'. However, this genus was accepted by the majority of botanists only recently when L. Garay and G.C.K. Dunsterville restated and defended the arguments in favour of transferring the five or so doubtful species from the *Miltonia* genus in their publication 'Venezuelan Orchids Illustrated' (Vol. 6, 1976). These species differ from *Miltonia* in respect of their single-leaved pseudobulbs and certain morphological peculiarities of their floral structure.

Whereas *Miltonia,* apart from the two exceptions just mentioned, is only represented in Brazil, the distribution area of this small genus of epiphytic and lithophytic orchids extends across Panama, Venezuela, Ecuador and Colombia.

115 Photo Ch. Weymuth

A particularly attractive species, from the olfactory standpoint, is

Miltoniopsis phalaenopsis,

which is found in Colombia, growing epiphytically in humid forests at altitudes of 1300–1600 m. Although *Miltoniopsis phalaenopsis* is not a typical moth-flower, it is more strongly-scented between midnight and dawn than during the day. This observation

Figures
113 *Miltonia regnellii* Rchb.f.
114 *Miltonia spectabilis* var. *moreliana*
115 *Miltonia schroederiana* (Rchb.f.) Veitch

supports Dodson's [11] supposition that this orchid, together with certain related *Miltoniopsis* species, are pollinated by crepuscular bees.

The scent of this enchanting orchid is soft and extraordinarily sensual, and is reminiscent of lily of the valley and cyclamen *(Cyclamen purpurascens)*. Analysis of the scent does in fact reveal a relationship with these two well-known 'scent flowers', as it possesses not only linalool, geranyl acetate and benzyl acetate, but also, more particularly, (E)-dihydrofarnesyl acetate (69), (E)-dihydrofarnesol (70) and γ-decalactone.

These substances are accompanied by a relatively large amount of elemicine (71), which gives the scent a discrete, yet characteristically spicy note (cf. analysis, p. 234).

A similar, though rather more rose-like scent is possessed by the no less attractive *Miltoniopsis roezlii*. Although easily perceived during the day, it appears to be at its strongest around dawn.

Odontoglossum H.B.K.

This large genus incorporates from 100 to 300 species, depending on the author consulted and the extent to which marginal groups have been hived off to form separate genera. Thus, for example, the famous *Odontoglossum*

116

grande and related species were recently allocated to the genus *Rossioglossum*. A feature common to all these species is that they only occur in the Neotropics, many species preferring the mountain regions of tropical and subtropical Central and South America.

An unmistakable species, with its impressive inflorescence, visible from afar, is

Odontoglossum cirrhosum.

Figures
116 *Miltoniopsis phalaenopsis* (Lind.et Rchb.f.) Garay et Dunsterv.
117 *Odontoglossum cirrhosum* Lindl.

Orchids of the American tropics

118

The overhanging inflorescence can grow to 60 cm in length, is densely packed with large individual flowers and is characterized particularly by narrow, tapering petals and sepals and brown to brownish-red spots. The scent of these orchids, which inhabit the humid mountain forests of Ecuador and Peru, is reminiscent of the familiar 'bois de rose', white-thorn and privet. This olfactory impression is based on the fresh-floral and diffusive accord of linalool and (E)-ocimene, which is given a distinctive character by the notes of benzaldehyde, 2-amino benzaldehyde and methyl cinnamate (cf. analysis, p. 236).

From the olfactory standpoint,

Odontoglossum constrictum

is very typical of a whole group of species of this genus. It grows in Venezuela and Colombia on the fringes of tropical mountain forests at altitudes of 1700–2400 m. The overhanging inflorescences, 15–20 cm in length, are in most cases (see Figure 118) also densely covered with rather small flowers that grow to about 4 cm. But *Odontoglossum constrictum* is subject to great variation, and also includes clones whose flowers are almost twice as large. Their floral scent is 'aromatic-spicy', strongly reminiscent of white-thorn and flower pollen, and the resinoid secretion of the leaves and branches of *Cistus* species, particularly *Cistus ladanifer.*

The high content of 2-amino benzaldehyde (72)—a widespread floral scent constituent, albeit as a minor component—is responsible for the white thorn and pollen aspects, whilst the herbal-balsamic, almost 'fougère'-like note is attributable to 3-phenylpropanal (73) and 3-phenylpropyl acetate (74) and the corresponding alcohol (cf. analysis, p. 236).

This complex of compounds 72–74, which we have previously encountered in a somewhat modified form in *Brassia verucosa* (p. 60), is a characteristic feature of the scent of a whole range of *Odontoglossum* species with similar morphology. These include, in particular, *O. cordatum, O. hallii* and *O. odoratum.*

In sharp contrast with the above-mentioned species, both in visual and olfactory terms, is

Odontoglossum pendulum,

which is also commonly listed under its synonym *O. citrosmum.* This plant creates an extremely attractive overall impression, and can be found on the Pacific side of Mexico at altitudes of 1400–2300 m in sunlit, mixed oak forests. As its other name suggests, its scent is rather reminiscent of lemon, particularly its aldehydic aspect. But initially one thinks of white-flowering *Trillium* species, of the sea, of dewy, fern-covered heathland, and perhaps also of damp hair—olfactory aspects

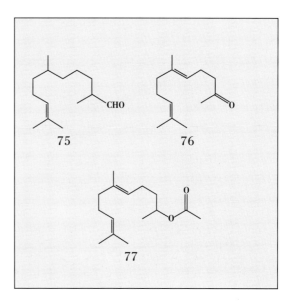

also found in the scent of 2,6,10-trimethyl-9-undecenal (75). This aldehyde is a familiar fragrance compound used in perfumery, specifically to reproduce the aforementioned impressions.

As the scent analysis described on page 236 shows, the main component of the scent of *O. pendulum* is a substance related structurally, and to a certain extent also olfactorily: (E)-6,10-dimethyl-5,9-undecadien-2-yl acetate, accompanied by smaller amounts of the corresponding alcohol.

These compounds were recently identified in the floral scents of *Zygogynum bicolor* and *Exospermum stipitatum* [39] (Winteraceae, both endemic to New Caledonia), and are derived from (E)-geranylacetone (76), another common substance also found in *O. pendulum.*

The last example of this genus represents another morphological group in which the usual 'rosy-floral' to 'aromatic-floral' scents tend to be in evidence. I refer to the delicate

Odontoglossum pulchellum,

which is a relatively common sight in mountain forests from Mexico to El Salvador. As

Figure

118 *Odontoglossum constrictum* Lindl.

Orchids of the American tropics

119

many as ten flowers grow on each 20–40 cm inflorescence. In contrast with other *Odontoglossum* species, they have not resupinated, and the lip still points upwards. Although *O. pulchellum* gives the impression of being a very consistent species, there appear to be various subspecies which differ particularly in the quantitative composition of their 'basic scent'.

All the plants evaluated contained geraniol as the predominant component, both in quantitative and olfactory terms. On the other hand, hydroquinone dimethyl ether, for example, can vary in its content between 10 and 60%, and methyl eugenol between 1 and 20%. Consequently, the 'rosy-floral' scent can seem either attractively fresh or rather dull (cf. analysis, p. 237).

Oncidium Sw.

With some 400–600 species, *Oncidium* forms one of the largest genera in the Neotropics, although its subdivision into clearly defined sections continues to pose problems, and its demarcation from the closely related genera *Odontoglossum* and *Miltonia* remains unclear. Consequently, certain marginal species have been allocated, more or less arbitrarily, to one or other genus.

Whereas most Oncidiums form pseudobulbs and live as epiphytes, certain species

120

have, for climatic reasons, not developed any such storage organs and grow terrestrially. In many Oncidiums, the basic flower colour is yellow, often with olive-brown to reddish-brown markings. Otherwise, white, pink, carmine-violet or dark brownish-red are the prevailing colours.

Although this species-rich family covers a broad spectrum of scents, it does have a number of typical olfactory characteristics. It is particularly noteworthy that the scent complex of 2-amino benzaldehyde plus 3-phenylpropyl alcohol and derivatives which characterizes the 'scent-shape-colour group' centred around *Odontoglossum constrictum* (p. 112) also occurs very commonly in yellowish-brown Oncidiums. Examples include *O. barbatum, O. divaricatum, O. gardneri, O. gravesianum* and *O. forbesii*. Another group of predominantly yellow oncidiums possesses 'aromatic-floral' scents that are often characterized by an aniseed or honey note. In some of these species, hints of 'ionone-floral' notes are discernible. Species in this group include *O. curcutum, O. micropogon, O. longipes, O. pusillum* and *O. sarcodes*.

Finally, the 'ionone-floral' aspect is very dominant in certain large-flowered, yellow-coloured species, such as *O. tigrinum* and *O. guianense*. A wide variety of scents is found in the pink to carmine-violet Oncidiums, ranging from 'rosy-floral' via 'spicy-floral' to phenolic.

After this rather cursory overview, four species, at least, of this diverse genus will be described in greater detail, starting with the epiphytic native of the Brazilian Serra dos Orgaos,

Oncidium longipes.

The sepals and petals of the flowers, which grow to about 3 cm in diameter, are brownish-red, whilst the large lip is a bright canary yellow.

Apart from the 'aromatic-floral' notes, its scent is strongly reminiscent of aniseed and cinnamon. This is due mainly to anis aldehyde, methyl anisate and cinnamic alcohol (cf. analysis, p. 237).

Belonging to the same olfactory group, and also a native of the Serra dos Orgaos in Brazil, is

Oncidium sarcodes.

The inflorescence reaches 100–150 cm in length and usually bears just a few flowers (though these are very large for the genus) with bright yellow lips and chestnut brown sepals and petals.

Its scent is strongly pervaded by ocimene, benzyl acetate and linalool, while relatively large amounts of (Z)-3-hexenol and the corresponding aldehyde lend this complex its characteristic green note (cf. analysis, p. 238). As with *O. longipes,* however, a clear aniseed note is also discernible.

The previously mentioned 'ionone-floral' aspect that typifies, to a greater or lesser extent, a whole range of yellow Oncidiums finds its highest aesthetic expression in

Oncidium tigrinum,

a species providing an unforgettable olfactory experience. This highly diffusive, completely transparent scent seems to invite one to

Figures
119 *Odontoglossum pendulum* (La Llave et Lex.) Batem.
120 *Odontoglossum pulchellum* Lindl.
121 *Oncidium longipes* Lindl. et Paxt.

Orchids of the American tropics

breathe it in deeply, and offers a perfect harmony of β-ionone and its derivatives plus ocimene and a range of minor components that are also of olfactory importance (cf. analysis, p. 238). We have already encountered similar 'ionone-rich' orchid scents in, for example, the genus *Encyclia* (p. 84).

In every respect a fascinating species, this orchid is a native of Mexico, growing at relatively high altitudes between 2000 and 2500 m. It has a particular affinity for oak trees, where it can form large clumps.

The same distribution area is also home to the very closely related *O. unguiculatum*, which is sometimes listed simply as a variety of *O. tigrinum*. In overall appearance, however, *O. unguiculatum* is rather larger, with certain differences in the fine structure of the lip, and those clones that I have been able to evaluate to date have proved to be scentless. This also applies to the very similar and very rare *O. splendidum*, although this species is a native of Guatemala and Honduras.

Figures
122 *Oncidium sarcodes* Lindl.
123 *Oncidium tigrinum* La Llave et Lex.
124 *Oncidium ornithorhynchum* H.B.K.

124

One of the best-known species of this genus is the ubiquitous

Oncidium ornithorhynchum,

without which no collection is complete. The species suffix *ornithorhynchum* refers to the shape of the column which, with a little imagination, resembles a bird's beak. *Oncidium ornithorhynchum* is found along the Pacific coast of Central America, from Mexico to Costa Rica. It develops overhanging, branched inflorescences, reaching up to 40 cm in length and covered with numerous, delightful pinkish-violet flowers measuring about 2 cm in diameter. Its scent is multifacetted, each individual facet being emphasized to a greater or lesser extent, depending on the time of day and the age of the flowers. But its essential character is provided by the main components neral, geranial, nerol and geraniol. Depending on the particular circumstances, the quantitative ratios of these compounds can be such that either the fresh-hesperidic or the 'rosy-floral' aspect predominates.

The combination of β-ionone, 2-amino benzaldehyde and vanilline can sometimes make the scent seem extremely sweet, whereas decanal, *trans*-2-nonenal and capric acid contribute an almost metallic aspect, reminiscent of *Himantoglossum hircinum* (cf. analysis, p. 237).

Of particular interest is the fact that the floral scent of *O. ornithorhynchum* also contains, as a minor component, the two diastereoisomers of chalcogran, otherwise known as biologically active compounds of the aggregation pheromone of the bark beetle *Pityogenes chalcographus* [40].

These two compounds are also contained, in rather greater quantities, in the scents of *Eria hyacinthoides* and *Liparis viridiflora,* two species presented in the next section.

Peristeria Hook.

This small genus incorporates only about six species which grow in the American tropics

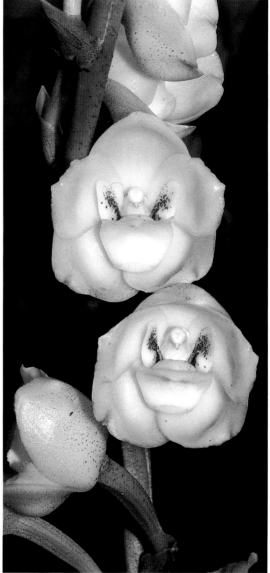

125

from Costa Rica down to Peru and Brazil. The genus name is derived from the Greek word 'peristerion' (dove), and refers to the dove-like structure formed by the fusion of the three-lobed lip with the column. The most famous of these 'dove orchids' is

Peristeria elata,

the national flower of Panama, popularly known as 'Flor del Espíritu Santo'. Unfortunately, its name and appearance led to its downfall. It is now a very rare species, particularly in Panama, where it used to grow in profusion. In contrast with most of the other *Peristeria* species, it grows terrestrially, especially in the transitional zone between grass savanna and forest. Leaves measuring up to a metre and more in length, resembling those of a young coconut palm, emerge from the 15–20 cm high, almost spherical, pseudobulbs. The inflorescence, which sprouts from the base, can easily reach one and a half metres in length, and bears 15–25 waxy, cream-white, strongly-scented flowers.

Nevertheless, its scent has a very simple structure, and the two main components, eucalyptol and phenylethyl acetate, are easily recognized by the human nose (cf. analysis, p. 238).

Rodriguezia Ruiz et Pavon

The 35 or so epiphytic species of this genus are distributed from Costa Rica, via Peru, to Brazil (most being found in Brazil). The small pseudobulbs bear one or two leaves, and the inflorescence emerges from the base to produce anything from a few to many medium-sized, often very attractively shaped and coloured flowers.

The two species most often seen in collections, *Rodriguezia decora* and *R. secunda*, I have invariably found to be scentless. All the more surprising, therefore, is the intense scent of the Venezuelan orchid

Rodriguezia refracta,

which emits almost pure (E,Z)-2,4-decadienal (31) (cf. analysis, p. 244), the compound that so strikingly typifies the scents of *Chondrorhyncha lendyana* (p. 52), *Pescatorea cerina* (p. 57), *Masdevallia tridens* (p. 100) and *Cattleya percivaliana* (p. 65). Apart from small amounts of the corresponding (E,E)-isomer, this scent—evoking a whole army of bugs—also contains 2.4% of a component that has not yet been fully elucidated. According to the mass spectrum and the retention time [41], the structure (E,Z,Z)-2,4,7-decatrienal may be proposed. This compound is known as an 'off-flavour' of various foods—including fish [42]—and, like the product in the orchid scent, produces a metallic 'codliver oil' smell at the outlet of the capillary column.

126   Photo Ch. Weymuth

Trichocentrum Poepp. et Endl.

This genus is related to *Rodriguezia* and covers about 12 rare species which also only occur in the American tropics. The best-known species is the eyecatching

Trichocentrum tigrinum,

This native of Ecuador provides another enchanting scent, a rather restrained, but ex-

Figures
125 *Peristeria elata* Hook.
126 *Rodriguezia refracta* Rchb. f.

tremely pleasant and full ylang-lily of the valley complex, based on a remarkable composition.

On the one hand is a very powerful ylang accord attributable to linalool, methyl benzoate and benzyl acetate while on the other hand this top note is subdued by a matrix reminiscent of damp petals and the heavier aspects of lily of the valley.

This matrix, which, in a slightly modified form, is also of crucial importance to the scent of *Rhynchostylis coelestis* (cf. p. 171), consists primarily of (E,E)-farnesal (78), (E,E)-farnesol (79), (E,E)-farnesyl acetate (80) and the corresponding (Z,E) isomers, together with (E)-dihydrofarnesyl acetate (69) and

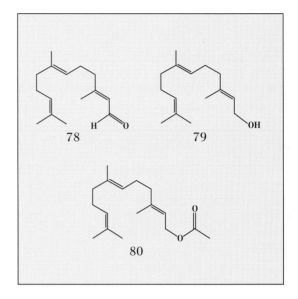

the corresponding alcohol (70) (cf. analysis, p. 244).

Zygopetalum Hook.

Various borderline species having been transferred to other genera, or assigned their own genera, *Zygopetalum* now has about 20 representatives, all of which grow in the tropical regions of South America. The inflorescence, emerging laterally from the pseudobulb, almost always bears brilliantly coloured, strongly-scented flowers.

Here we shall look at

Zygopetalum crinitum,

an inhabitant of eastern Brazil. Its flowers are distinguished by the crown-like arrangement of the petals and sepals, and the large white lip, which is traversed by purplish-red veins.

Its scent has a special 'spicy-floral' character, reminiscent of certain aspects of the narcissus, and creates a remarkably fresh

Photo Ch. Weymuth

Figures
127 *Trichocentrum tigrinum* Lindl. et Rchb.f.
128 *Zygopetalum crinitum* Lodd.

impression. The essential character of this almost indescribable scent is largely attributable to the interplay of estragole, geranial, geraniol and its acetate, eugenol, methyl anthranilate and a relatively large quantity of (Z,E)-farnesal and (E,E)-farnesal, embedded in plenty of nerolidol (cf. analysis, p. 246).

The closely related *Zygopetalum mackaii* has a scent with a much simpler structure that is strongly reminiscent of hyacinth and narcissus.

Orchids of the African tropics

Angraecoid orchids

As already mentioned, the orchid flora of Africa—including Madagascar, which itself makes a significant contribution—forms a uniform complex specifically characterized by a high proportion of white-flowering, night-scented species (pp. 15, 29). Proceeding on the assumption that this geobotanical area accommodates about 10% of all orchid species [43], and that approximately 8% of all orchid species demonstrate the syndrome of moth pollination [11], one can conclude that Africa, including Madagascar, must be home to over 1200 representatives of this floral ecological group. The large majority of these belong to genera of the subtribe *Vandaeae* and, taken together, are allocated to the group known as angraecoid orchids. The term 'angraecoid' derives from the largest genus *Angraecum,* which encompasses about 200 species. *Angraecum,* in turn, is the latinized version of 'anggrek', the Malay word for orchid.

For the 'non-taxonomist', the pronounced nectar-containing spur is the most obvious feature of these basically white flowers, which may also be tinged with pink, green or yellow. Naturally, other genera also possess spur-carrying flowers, but in the angraecoid orchids, the spur is at least equal to the diameter of the flower. Representatives of this group do not possess pseudobulbs, but are typically 'monopodial'. In other words, the main stem grows by the continual redevelopment of its terminal bud, with flower stalks emerging from the leaf axes. The evergreen leaves of these (mostly epiphytic) plants are usually asymmetrically bilobed. All angraecoid orchids show a common floral structure that is ideally adapted to moth pollination. They possess two pollinia connected to a stipe with the sticky disk known as the viscidium.

Particularly fascinating examples of these magical nocturnal beings are found in the genus

Aerangis Rchb.f.

Following the revision by J. Steward (Kew Bulletin 1979) of this genus, which was established by Reichenbach (1865), *Aerangis* now incorporates about 50 species which occur in the rainforests and mountain and savanna forests of tropical Africa and Madagascar. As the following examples also illustrate, the flowers of all *Aerangis* species have a basic white colouring, with occasional greenish to salmon-pink tinges, and invariably possess, in relation to their flower size, long or very long spurs. Of particular interest in the context of our main theme is the fact that, under natural conditions, all species are practically scentless during the day. They only begin to release their scent at sunset, and in all the specimens that I have personally encountered, emanation reaches a peak between 21.30 and 23.00 hr.

The distribution area of the first representative,

Aerangis appendiculata,

stretches from Tanzania, via Mozambique and Malawi down to Zimbabwe, and grows at altitudes around 2000 m in evergreen forests. This relatively small-flowered species bears a 4–6 cm spur, and is very similar in flower type and scent (excepting the spur, which is only half as long) to *A. somalensis* (p. 129). Its pronounced 'white-floral' scent is attributable mainly to linalool and methyl benzoate, and is characterized by p-cresol, eugenol, indole and vanilline. The animalic note of p-cresol is particularly noticeable during the first few hours; subsequently it diminishes and be-

comes fully integrated in the pleasant scent complex (cf. analysis, p. 188).

Aerangis biloba

is a widespread species which grows in West Africa from Senegal right across to Nigeria and Cameroon. It was probably the first species of this genus to flower in a European collection. The inflorescence can sometimes be very impressive, carrying up to 20 flowers. As often occurs with *Aerangis* species, the length of the spur is very variable, ranging from 4.5 to 7.5 cm. Its powerful scent has a pronounced 'white-floral' aroma, with methyl benzoate clearly in evidence during the first half of the night. From midnight until morning, the scent seems much more rounded, being reminiscent of lily and tiaré *(Gardenia taitensis)* (cf. analysis, p. 188).

A much more complex scent—although it could also be summed up simply as aromatic 'white-floral'—is possessed by the exceptionally elegant

Aerangis brachycarpa,

whose spur can reach up to 19 cm in length.

Its jasmine-like basic skeleton is formed from linalool, benzyl acetate, benzyl alcohol, phenylethyl alcohol and indole, with additional contributions from anethole (aniseed-like), methyl phenylacetate together with vanilline (sweet, honey-like) and p-methyl anisole (ylang-ylang aspects) (cf. analysis, p. 189).

Aerangis brachycarpa has an exceptionally large distribution area, extending from Ethiopia via Kenya, southwards to Zambia and Angola. It is one of the more common *Aerangis* species, and is found particularly growing alongside rivers and streams where a uniform microclimate can develop.

Until recently, much confusion surrounded another species growing in the highlands of Kenya and northern Tanzania, often together with *A. brachycarpa*. It had never been described scientifically, and could occasionally be found in collections under the name of *A. brachycarpa* or its synonym *A. friesiorum*. In her revision of the African species (Kew Bulletin 1979), J. Stewart acknowledged the reason for this confusion and, accordingly, gave this familiar species the name

Aerangis confusa.

I, too, have been confused by this enchanting species, since it can be very variable, with marked differences also in its scent composition. The basic character always remains the same, however, and is reminiscent of tuberose and gardenia, aspects that are even more pronounced in the scent of *Aerangis kirkii* (p. 128).

The same lactone is responsible in both cases: *cis*-4-methyl-5-decanolide (81), a substance that has yet to be described in the literature. Since we have so far only managed to identify this olfactorily interesting lactone in the two above-mentioned *Aerangis* species, we would like to name it aerangis lactone. This compound (81) represents 2–5% and 20–30%, respectively, of the floral scent of *A. confusa* and *A. kirkii* adsorbed onto activated charcoal.

Interestingly, aerangis lactone is accompanied in both cases by methyl 3-methyl octanoate (82), a substance that has also only been detected hitherto in these two orchid scents. This methyl ester, which is strongly reminiscent of overripe fruit and, to a certain extent, wine yeast, could possibly have the same biological precursor as aerangis lactone. More of structural than ol-

131

factory interest is the C_{16}-homoterpene (83), accounting for about 30% of the scent of *A. confusa*. As Boland [44] recently confirmed by experiment, this substance must be considered to be a metabolite of geranyllinalool. Over the course of the last seven years, we have detected this tetraene (83) in the scents of 35 orchid species and in those of an equal number of other flowering plants [16].

Figures
129 *Aerangis appendiculata* (De Wild.) Schltr.
130 *Aerangis biloba* (Lindl.) Schltr.
131 *Aerangis brachycarpa* (A. Rich.) Dur. et Schinz

Orchids of the African tropics

132

Orchids of the African tropics

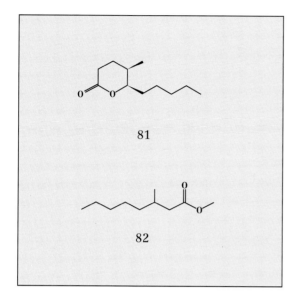

Aerangis distincta,

was described just five years ago (Kew Bulletin 1987), again by J. Stewart, a great authority on the angraecoid orchids. This very rare species is a native of Malawi and particularly likes to grow near rivers at altitudes between 650 and 1800 m. In overall appearance it perhaps most closely resembles *A. brachycarpa,* although it differs 'distinctly' also in its floral scent. It produces a pronounced 'spicy white-floral' effect, is reminiscent of certain lily and *Nicotiana* species, and is based largely on a complex of methyl salicylate, phenylethyl alcohol, a relatively large quantity of indole and benzyl salicylate, ultimately deriving its warm, spicy character from a large amount of isoeugenol and vanilline.

In both *A. confusa* and *A. kotschyana* (p. 129), this compound (83), serving as a main component, is accompanied by the structurally related methyl esters 84–86, which also probably represent metabolites of geranyllinalool. These compounds 84–86 could prove to be taxonomically significant within the variable species *Aerangis confusa*. Amazingly enough, another clone also native to Kenya possesses a pattern of scent constituents, which, apart from the complete absence of the methyl esters 84–86 is almost identical (cf. analysis, p. 189).

One of the most recent discoveries in this genus,

Although this scent complex is based solely on well-known perfumery products, it is unique amongst the 250 orchid scents investigated, and it is difficult to believe that the even simpler combination of methyl salicylate, phenylethyl alcohol and isoeugenol only occurs in its closest relative, *Aerangis brachycarpa* (cf. analysis, p. 191).

Aerangis fastuosa

is an epiphyte endemic to the eastern part of

Figure
132 *Aerangis confusa* J. Stewart

127

Orchids of the African tropics

133

134

Madagascar which develops pure white flowers that are very large in relation to the plant and have a decidedly neat and tidy appearance. Each flower has a spur 7–8 cm in length. During the night, it emanates an almost clinically pure, 'white-floral' scent, reminiscent of the lily *(Lilium longiflorum)*. This scent is also based on familiar floral scent compounds, particularly on linalool, methyl salicylate and homologues, benzyl alcohol and p-cresol which, together, already account for more than 80% of the composition. However, the quantitative ratios are such that this scent also appears to be characteristic of the species (cf. analysis, p.191).

Producing a much warmer and sensual effect, the scent of

Aerangis kirkii

is reminiscent of aspects of tuberose and gardenia, and is closely related, in olfactory terms, to that of *Aerangis confusa*. As already mentioned, it contains very large amounts of aerangis lactone (81), again accompanied by methyl 3-methyloctanoate (82). The scent of *A. kirkii* also contains another lactone which is structurally related to compounds 81 and 82, namely *cis*-3-methyl-4-decanolide, which has also never been identified as a natural substance (cf. analysis, p. 191). A notable difference, however, is that the metabolites of geranyllinalool (83–86) detected in *A. confusa* and *A. kotschyana* are lacking in *A. kirkii*.

Aerangis kirkii is a shade-loving epiphyte of hot coastal forests of Kenya, Tanzania and Mozambique.

One of the most magnificent of African orchids is

Aerangis kotschyana,

which has an enormous distribution area covering the whole of tropical Africa and southern Africa down to the Transvaal. The lip, which is considerably broader than the sepals and petals, gives the flower its characteristic appearance. But its most astonishing feature is its long spur, measuring as much as 23 cm, which invariably incorporates one or more corkscrew turns in its lower half. The strong scent of these flowers is reminiscent of certain jasmine species that contain only very small amounts of methyl anthranilate but large amounts of vanilline. Of particular importance to this olfactory impression is the interaction of linalool, benzyl acetate, cinnamic aldehyde, indole and vanilline. The structural features of particular note are the previously discussed metabolites of geranyllinalool (83–86) (cf. analysis, p. 192).

The 'white-floral' scent bouquet of the *Aerangis* species is completed by

Aerangis somalensis,

136   Photo Ch. Weymuth

a species growing mainly in Somalia, Kenya and Ethiopia in fairly arid, stony areas. The 10–20 cm long inflorescence bears from 5–20 relatively small flowers, each with a conspicuous spur reaching 10–15 cm in length. Overall, the flowers perhaps most closely resemble those of the smaller *A. appendiculata*. Its very intense scent produces a rather sim-

135

Figures
133 *Aerangis distincta* J. Stewart
134 *Aerangis fastuosa* (Rchb.f.) Schltr.
135 *Aerangis kirkii* (Rolfe) Schltr.
136 *Aerangis kotschyana* (Rchb.f.) Schltr.

Orchids of the African tropics

137

ple 'white-floral' accord, the main components of which are linalool, methyl benzoate and nerolidol (cf. analysis, p. 193).

Aeranthes Lindl.

The 40 or so species of the genus *Aeranthes* grow on Madagascar and the surrounding islands, and manage to combine, in a very special way, floral splendour with bizarre, spidery and insect-like forms. The often translucent, greenish to greenish-white flowers generally hang from long, threadlike inflorescences and could, as their name suggests, easily be thought of as 'air flowers'. The genus was first established by Lindley in the 'Botanical Register' (1824), with

Aeranthes grandiflora.

This rare plant lives in the hot, humid forests of central and eastern Madagascar and, contrary to claims in the literature, its flowers are anything but scentless. As might be expected from the floral structure, *Aeranthes* species are night-scented, with a culinary twist in the case of *A. grandiflora*, since its scent is initially reminiscent of freshly-baked butter cookies and, thereafter, of floral-herbaceous notes.

This aspect, unique amongst all the orchids investigated, is largely due to 2,3-pentan-

138   Photo Ch. Weymuth

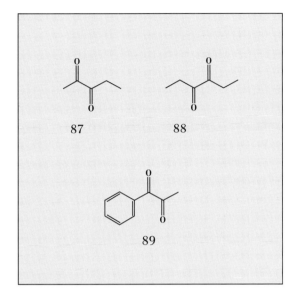

dione (87), 3,4-hexandione (88), phenyl-1,2-propandione (89) and benzaldehyde (cf. analysis, p. 193).

Angraecopsis Kraenzl.

Following its revision by Summerhayes in 1951, this small genus, a close relation of *Aerangis* and *Diaphananthe,* still incorporates 10 species that occur in tropical Africa and Madagascar. These are all small epiphytes with long roots and often only a few leaves and pendulous racemes up to 10 cm long. The latter are often covered with numerous, very small greenish to creamy-white translucent flowers that are scentless during the day and are easily overlooked. After sunset, however, their very strong scent renders them much more conspicuous.

An impressive example is the epiphytic inhabitant of the mountain forests of Kenya, Tanzania and Mozambique and adjacent areas,

Angraecopsis amaniensis,

whose flowers, though measuring just a few millimetres in diameter, can emit a scent as strong as that of the large-flowered angraecoid orchids. It is not exactly an elegant perfume, but rather a garish, fruity, green-floral accord that would have no difficulty in attracting the pollinators to the flower, even on a dark moonless night.

This accord is based primarily on (Z)-3-hexenyl acetate, methyl benzoate, the main component benzyl acetate, geraniol and surprisingly large quantities of methyl 2-methoxybenzoate (cf. analysis, p. 195).

The closely related *Angraecopsis breviloba* possesses a very similar scent composition.

Figures
137 *Aerangis somalensis* (Schltr.) Schltr.
138 *Aeranthes grandiflora* Lindl.
139 *Angraecopsis amaniensis* Summerh.

Orchids of the African tropics

Angraecum Bory

The *Angraecum* genus incorporates about 200 species that grow in tropical and southern Africa, as well as Madagascar and the surrounding islands. These include both dwarf orchids, measuring just a few centimetres tall and with floral diameters of perhaps 5 mm, and 2–3 metre high giants with flowers attaining over 20 cm in diameter. Yet they all create the typical angraecoid effect, only perceived elsewhere in the other angraecoid orchids and *Neofinetia falcata*.

This genus was first established as early as 1804 by the French travelling researcher Bory de Saint Vincent, who described

A. eburneum in his book 'Voyages des quatre principales isles des mers d'Afrique'.

Its first representative,

Angraecum aporoides,

is closely related to the rather smaller *A. distichum*. This species is relatively common in orchid collections and, like *A. distichum,* is a native of southeast Africa. Its scent is markedly 'aromatic white-floral' and strongly characterized by anis aldehyde and benzaldehyde.

The scent analysis described on page 195 is based on a scent sample obtained from a single specimen of these flowers, which only grow to about 7 or 8 mm, during the period from 18.00–22.00 hr.

In marked contrast with this small-flowered species is

Angraecum bosseri,

an orchid that is endemic to Madagascar and that was first described at the beginning of the 1970s. It is closely related to the famous *Angraecum sesquipedale,* the 'Star of Madagascar', which was presented earlier as a typical example of a 'moth-flower' (p. 29). Overall, *A. bosseri* is rather smaller and more delicate, the inflorescence is invariably unifloral (*A. sesquipedale* has three or four flowers) and the spur of the flowers—which are also slightly smaller—'only' reaches 25 cm, as against the record length of 35 cm. There are, however, highly characteristic differences between the scents of these two species, both from the olfactory and the chemical point of view (analysis *A. bosseri,* p. 195, *A. sesquipedale,* p. 197). The scent of *A. bosseri* is much more brilliant and transparent, and is unquestionably 'spicy white-floral'.

The main components of this scent are linalool and phenylethyl alcohol, though it derives its special character from fairly large quantities of eugenol, isoeugenol, chavicol and isochavicol, a unique combination amongst the orchids investigated. However, the oximes of isovaleraldehyde, 2-methylbutyraldehyde and phenylacetaldehyde that are typical of many *Angraecum* species (p. 30) are only present in fairly small quantities. On

Orchids of the African tropics

141

the other hand, it is these specific oximes that are probably of taxonomic significance for

Angraecum eburneum,

which is currently divided into four subspecies. Accordingly, the floral scent of *A. eburneum* ssp. *eburneum,* a native of Réunion, is very rich in isovaleraldoxime (3, p. 30), whilst that of ssp. *superbum* contains large quantities of 2-methylbutyral-doxime (2, p. 30). This subspecies mainly grows epiphytically in evergreen coastal forests on the Comoros, the Seychelles and the east coast of Madagascar.

As can be seen from Figures 142 and 143, the flowers of *A. eburneum* ssp. *eburneum* are significantly smaller than those of ssp. *superbum.* Moreover, the petals spread straight out, the lateral sepals are not concealed beneath the lip, and the lip itself has a rather rounded outline.

The scent of the type-species, ssp. *eburneum,* is exceptionally strong, indeed almost offensively 'aromatic white-floral'. It is based largely on benzyl acetate, methyl benzoate, benzaldehyde, phenylacetaldehyde, phenylethyl alcohol, indole and the previously mentioned isovaleraldoxime, which is released in large quantities between sunset and 22.00 hr, lending the scent an astringent note during this period (cf. analysis, p. 196). Interestingly, the content of isovaleraldoxime decreases as the content of phenylacetal-

142

dehyde increases, the latter ultimately accounting for about 15% of the headspace sample after midnight. At this point, the scent is extremely powerful, almost pungent, and reminiscent of aspects of hyacinth. During the early evening hours, the scent of ssp. *superbum* produces a very astringent, herbaceous-green, almost narcotic effect, due to

Figures

140 *Angraecum aporoides* Summerh.
141 *Angraecum bosseri* Senghas
142 *Angraecum eburneum* ssp. *eburneum* Bory

Orchids of the African tropics

143

144

the high content of (Z)-3-hexenol and 2-methylbutyraldoxime, and subsequently develops into a lily-like accord after midnight (cf. analysis, p. 196).

By contrast, a pronounced but familiar 'white-floral' effect is produced by the scent of

Angraecum eichlerianum.

This very strong scent is likewise only emitted during the night, and is particularly evocative of jasmine and lily *(Lilium grandiflorum)*. The basic accord, which is composed of linalool, methyl benzoate, methyl salicylate and indole, can very easily be extended in the direction of jasmine or lily (cf. analysis, p. 196).

A. eichlerianum is one of the few species in this genus that occurs in tropical west Africa. It develops relatively long stems, which bear numerous leaves arranged in two lines and which usually climb up tree trunks. The single flowers reach a diameter of 8 cm and possess a particularly striking upright spur with a funnel-shaped base. Similarly shaped and scented flowers are produced by *A. infundibulare,* a species found in the same distribution area.

Diaphananthe Schltr.

The very name draws attention to the diaphanous, delicate, usually white to pale green flowers, which grow in profusion on inflorescences measuring as much as 40 cm or more in length. This genus is currently thought to total about 50 species distributed across the whole of tropical Africa and southwards down to Natal.

The most famous of these beautiful species possesses the melodious name of

Diaphananthe pellucida.

Figures
143 *Angraecum eburneum* ssp. *superbum* Bory
144 *Angraecum eichlerianum* Kraenzl.
145 *Diaphananthe pellucida* (Lindl.) Schltr.

145

Orchids of the African tropics

Upwards of 40 of these translucent, almost crystalline flowers adorn the inflorescence, which can be up to 50 cm in length. During the night, they emanate a very curious, herbaceous-green and, to some extent, heavy-floral and sweet scent. Perhaps it most closely resembles 'brouts absolue des eaux', a perfumery specialty obtained by extraction of the distillation water produced during the preparation of neroli essence.

Analysis of the scent sample showed that the oximes 2–4 (p. 30), which are commonly found in angraecoid orchids, are of particular importance here as well. But in this case the scent is dominated by the nitriles 90–92 arising from the oximes and particularly by phenylacetonitrile (92), which accounts for about 60% of the overall scent (cf. analysis, p. 218).

Considerably smaller, but no less fascinating in appearance, are the flowers of

Diaphananthe pulchella,

whose habitat extends from Uganda, via Kenya and Tanzania, to Malawi. Its strong scent has a pronounced 'aromatic white-floral' character, with an additional pleasant, woody note, and is largely characterized by methyl benzoate, methyl salicylate, benzyl acetate and nerolidol (cf. analysis, p. 219).

Plectrelminthus Raf.

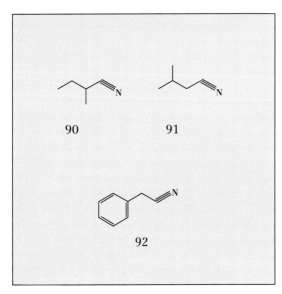

One of the most extraordinary of all the angraecoid orchids was originally described by Lindley as *Angraecum caudatus,* but was subsequently removed from the *Angraecum* genus by Summerhayes owing to its highly characteristic column structure and

---

Figures
146 *Diaphananthe pulchella* Summerh.
147 *Plectrelminthus caudatus* (Lindl.) Summerh.

Orchids of the African tropics

147

other morphological features; it thus became the sole representative of a separate genus,

Plectrelminthus caudatus.

This large epiphyte is encountered with relative frequency in tropical West Africa, particularly in Sierra Leone, although it is still rarely seen in collections.

The inflorescence, up to 60 cm in length, carries 5–10 large non-resupinated flowers, each bearing a corkscrew spur up to 20 cm long. The olive-green sepals and petals form a striking contrast with the pure white lip. At night, this unusual looking flower emits a scent that is probably unique in the whole realm of flowering plants. It is strongly reminiscent of a mossy, fungus-covered forest floor. The 'white-floral' accord that is typical of nocturnal scents is pushed right into the background, and is, in fact, only discerned by scent analysis.

This unique scent can be attributed to the combination of numerous esters, especially those of tiglic acid, of which hexyl tiglate (93) dominates in both quantitative (58%) and olfactory respects. A particularly characteristic contribution is also provided by tiglyl tiglate (94), a substance which we first identified 15 years ago in the scent of *Gardenia jasminoides* [45] and which accounts for about 7% of the composition (cf. analysis, p. 241).

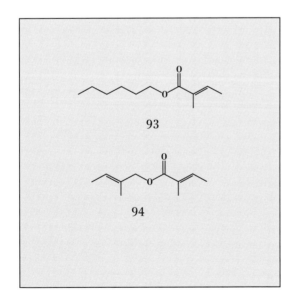

Rangaeris (Schltr.) Summerh.

After the 'odeur sousbois' of *Plectrelminthus caudatus,* it seems particularly appropriate to round off the group of angraecoid orchids with a representative of the genus *Rangaeris* (a near anagram of *Aerangis*), which only includes about five species. Until it was classified as a separate genus (Summerhayes, 1933), *Rangaeris* was considered as a section of *Aerangis.*

A species fairly often encountered in collections is

Rangaeris amaniensis,

which likes to grow epiphytically on *Juni-*

perus species in the mountain forests of Uganda, Kenya and Tanzania. The white flowers, very similar in overall appearance to *Aerangis* species, possess spurs up to 15 cm long, and emit a very pleasant, characteristic 'white-floral' scent that perhaps most closely resembles a combination of the scents of *Stephanotis,* lily of the valley and lily. It is based on methyl salicylate, phenylethyl alcohol, methyl cinnamate, farnesol and linalool, all very familiar scent compounds (cf. analysis, p. 243). But it is the quantitative aspects of this basic accord that make this scent unique amongst all the floral scents investigated to date.

Although 'white' predominates in the African orchid flora in both visual and olfactory respects, this continent is also home to a number of attractively coloured day-scented representatives with notable scents, as the following examples will show.

Ancistrochilus Rolfe

This genus also occurs only in Africa, and incorporates just two, albeit magnificently coloured, epiphytic species, with a distribution area from West Africa to Uganda.

Ancistrochilus rothschildianus

is a species occasionally seen, and much admired, in orchid collections. The flowers—which are large in relation to the plant—grow to 5 cm in diameter, are a delicate lilac-pink to purple in colour and emit a strong, waxy, almost metallic scent, rather reminiscent of wine yeast. This unusual scent derives from a series of straight-chain methyl and ethyl esters plus their free acids (which are sometimes present in smaller amounts). Ethyl caprinate and ethyl laurate are the dominant components (cf. analysis, p. 194).

These esters are accompanied by fairly large quantities of 1-pentadecene, (Z)-1,8-pentadecadiene and a related triene with a molecular weight of 206, probably (Z,Z)-1,8,11-pentadecatriene. These olfactorily less significant C_{15}-hydrocarbons are known to be constituents of Echinacea species [e.g. 46].

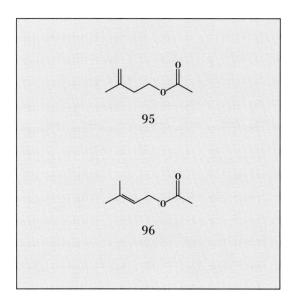

Polystachya Hook.

The *Polystachya* genus incorporates about 200 small to medium-sized epiphytic (or, in a few cases, lithophytic or terrestrial) species, most of which grow in Africa. Depending on the climatic conditions of the respective distribution areas, the shoots growing on the short rhizomes may appear bulb-like due to thickening of the stem. The terminal inflorescence contains anything from a few to a large number of (mostly small) non-resupinated flowers. At first glance, the species of this large genus are certainly not as spectacular as the large-flowered Cattleyas, but should you care to look more closely, you will find them similarly fascinating. If you had not already noticed it, you will then realize that many of these species are also scented and that their scents are remarkably diverse. For example, the species illustrated,

Polystachya campyloglossa,

emanates a very characteristic, fresh green-herbaceous scent reminiscent of bananas and strawberries, and containing as its main com-

Figure
148 *Rangaeris amaniensis* (Kraenzl.) Summerh.

Orchids of the African tropics

149

ponents isoprenyl acetate (95), prenyl acetate (96) and the corresponding alcohols. Smaller, though olfactorily important, components include dihydro-β-ionone and β-ionone (cf. analysis, p. 241).

This species can vary greatly in its colouring, and grows from Uganda, via Kenya to Tanzania and Malawi at altitudes between 1100 and 2700 m.

Another highly variable species is

Polystachya cultriformis,

whose small flowers can be white, pink, yellow or yellowish-green. Its distribution area is very large, covering the whole of tropical Africa, including Madagascar. The scent of the yellow-flowering type (illustrated here) is particularly pleasant, evoking lily of the valley and lime blossom. These olfactory relationships are also apparent at the chemical level. The 'muguet complex', the main components of which are nerolidol, dihydrofarnesol and farnesol, is enhanced by a range of minor components, including γ-decalactone, 2-amino benzaldehyde, β-ionone and methyl jasmonate, to produce the typical lime blossom notes (cf. analysis, p. 242).

A small-flowered species growing in the rainforests of western Uganda and eastern Zaire,

Polystachya fallax,

makes its presence felt by an even more intense scent. Its flowers likewise grow to just 5 to 7 mm in diameter, and its scent, which is rich in methyl anthranilate and indole, is reminiscent of narcissus and sampaquita *(Jasminum sambac).* A note of tropical fruits is also discernible, originating principally from numerous esters. Various bifunctional esters which, like the oximes discussed above, are derived from amino acids, are of both olfactory and structural interest.

Particularly noteworthy are the two diastereoisomers of methyl 2-hydroxy-3-methylvalerate (97), methyl 2-hydroxy-4-methylvalerate (98) and the keto ester (99) derived from 97 (cf. analysis, p. 242). During the past ten years we have identified esters 97

150

and 98 and their derivatives in various floral scents, including *Boronia megastigma, Michelia champaca* and *Stephanotis floribunda* [47]. As has recently been reported, the two diastereoisomers of 97 and the corresponding acetates play an important role in the pollination of *Eupomatia* species [48].

Figures
149 *Ancistrochilus rothschildianus* O'Brien
150 *Polystachya campyloglossa* Rolfe

Orchids of the African tropics

151

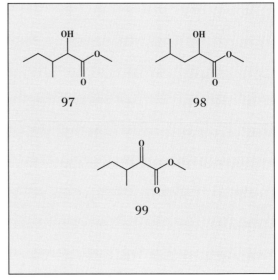

The last representative of this genus, which will conclude this overview of the African orchids, is

Polystachya mazumbaiensis,

a rare plant known only to occur in the Usambara Mountains (Tanzania, Mazumbai Forest Reserve), and first described just a few years ago by Cribb and Podzorski in the Kew Bulletin (1978). This small orchid grows epiphytically or lithophytically at altitudes between 1150 and 1450 m, and develops five to ten small flowers on inflorescences just 5 cm long. These flowers are tinged with pink-violet hues and, once again, emit a very characteristic scent. Its basic character is 'aromatic

152

153

rosy-floral', accompanied by 'ionone-floral' and herbaceous-green aspects. In chemical terms this translates as a combination of geranylacetone (the main component) plus anisyl alcohol, including derivates, cinnamic aldehyde, 2-N-methylamino benzaldehyde and dihydro-β-ionone; it derives its green-herbaceous facet from jasmone, methyl (Z)-4-decenoate and (Z)-3-nonen-1-ol (cf. analysis, p. 243).

Figures
151 *Polystachya cultriformis* (Thou.) Sprengel
152 *Polystachya fallax* Kraenzl.
153 *Polystachia mazumbaiensis* Cribb et Podzorski

Orchids of the Indo-Australian tropics and subtropics

Aerides Lour.

The genus *Aerides* encompasses twenty epiphytic species which inhabit a belt stretching from the Indian peninsula to the Moluccas and across the Philippines to southern Japan. Like those of the previously discussed genera *Aerangis* and *Angraecum*, the *Aerides* species favour a monopodial type of growth, i.e. upwards and not laterally. The (mostly pendulous) racemes are covered, to varying degrees of profusion, with numerous eyecatching, fleshy or waxy flowers. Most species emit strong and often pleasant scents, generally based on 'rosy-floral' to 'aromatic-floral' accords. This genus was established by the Portuguese missionary and botanist J. Loureiro (1790) who, alluding to their epiphytic lifestyle, referred to these splendid plants as 'daughters of the air', and accordingly named them *Aerides*.

One very attractive species, which inhabits southern Burma and Thailand, is

Aerides crassifolia.

Pendulous racemes, tightly packed with lilac-red flowers, grow from the axes of the coarse, rather fleshy leaves. The flowers grow to 4–5 cm in diameter and are characterized mainly by the complex shape of the lip—which has a forward-leaning spur—and by its scent. The latter, however, produces a rather simple 'aromatic-floral' effect, being strongly dominated by benzyl acetate and essentially containing only aromatic compounds of this type (cf. analysis, p. 193).

An extremely pleasant scent, on the other hand, is given off by the white and amethyst-purple flowers of

Aerides fieldingii.

It is reminiscent of lily of the valley and cyclamen, and is based on a 'muguet complex' consisting of nerolidol, dihydrofarnesol, farnesol and farnesal. We have already encountered this complex, in slightly modified form, in *Miltoniopsis phalaenopsis* (p. 109) and *Polystachya cultriformis* (p. 141). A number of compounds—which are present in smaller quantities and include phenylethyl acetate, linalool and ocimene—give the scent its specific character. (cf. analysis, p. 194).

The inflorescence of this native of the tropical Himalayas can reach up to 60 cm

154

155

156

in length, and bears a large number of wonderfully-scented flowers, each measuring about 4 cm across. In morphological and, to a certain extent, olfactory terms, *A. fieldingii* resembles *A. multiflora* Roxb., a species growing in the same habitat. Its 'aromatic rosy-floral' scent is, however, rather simpler and is quite strongly influenced by phenylethyl acetate.

Perhaps the most beautiful of these 'daughters of the air' is

Aerides lawrenceae,

a native of the Philippine island of Mindanao. The uniquely shaped flowers form tightly packed clusters on pendulous racemes. In the illustrated specimen, sepals, petals and lips have a most attractive carmine-pink tinge. A particularly characteristic feature is the large, forward-leaning, funnel-shaped spur emerging from the lip.

The scent of these flowers is also unique and unmistakable, providing an unforgettable experience. It is a 'rosy-floral' scent based on large quantities of nerol and geraniol and

Figures
154 *Aerides crassifolia* Par. et Rchb.f.
155 *Aerides fieldingii* Jennings
156 *Aerides lawrenceae* Rchb.f.

producing, on the one hand, a very sweet and honey-like effect on account of its relatively high content of methyl phenylacetate and, on the other, a pleasant freshness by way of neral, geranial and farnesal. Methyl anthranilate and indole, though present in fairly small quantities, form a very striking contrast with the 'rosy-floral' basic scent (cf. analysis, p. 194).

Bulbophyllum Thou. and Cirrhopetalum Lindl.

There will always be certain plant groups whose classification, whether as sections of a genus or as independent genera, is controversial. This applies particularly to *Bulbophyllum* which, with over 1000 species, is the largest genus in the orchid family. All species grow epiphytically, and their principal distribution area is Southeast Asia. A sizable number of *Bulbophyllum,* however, also occur in South America, Australia and eastern Africa, including Madagascar. In contrast to the mainly unifloral representatives, at least 50 species possess a characteristic arrangement of several flowers forming a more or less semicircular umbel or fan. Moreover, the two lateral sepals in these species are very long and frequently curled (cirrhus petalon), hence Lindley's choice of *Cirrhopetalum* for the genus name. Since, however, many transitional forms exist and no demarcation is possible, various botanists believe that *Cirrhopetalum* should be fully incorporated within *Bulbophyllum.*

A feature common to both genera is the highly mobile lip, which resembles a seesaw. From the pollination standpoint, this may be considered as characteristic of the genus. A fly or midge landing on the lip tips over and falls against the column, thus acting as a pollinator. In order to attract the appropriate insects, such flowers are often reddish-brown to yellowish-brown in colour and emit carrion-like or fecal scents (cf. p. 31).

The first representative to be described,

Bulbophyllum lobbii,

157

has become a favourite among collectors because of its attractive flowers. Even Lindley considered this plant to be the most beautiful orchid in this huge genus.

The flowers grow up to 10 cm in diameter and emit a rather pleasant—though perhaps somewhat strong —scent which is reminiscent of aspects of jasmine and orange blossom and is in no way comparable to one of the notorious *Bulbophyllum* smells. It is based largely on ocimene, linalool and derivatives of these two compounds, relatively large amounts of phenylacetonitrile and indole. Another main component, (E)-4,8-dimethyl-1,3,7-nonatriene (100), also makes its own contribution, with a note located somewhere between limonene and (E)-ocimene. This C_{11}-homoterpene (100), usually accompanied by its (Z)-isomer 101 in the ratio 50:1 to 100:1, shows a distribution comparable with that of C_{16}-homoterpene 83 (cf. p.127), and is probably produced from nerolidol via a similar biogenesis as 83. The scent of *Bulbophyllum lobbii* also contains, as a derivative of 100, 0.6–1.0% (E)-2(3)-epoxy-2,6-dimethyl-6,8-nonadiene (102), which we discovered several years ago as a new natural substance in the scent of the so-called Queen of the night (*Selenicereus hamatus,* Cactaceae) [16]. Equally noteworthy is the presence of the two possible epoxides of (E)-ocimene, together with isovaleraldoxime (E:Z ~ 2:1) and phenylacetaldoxime (cf. analysis, p. 201).

In no other species are the contrasts within this vast genus more in evidence than in

Bulbophyllum barbigerum.

Although, unlike the majority of the species, it is a native of West Africa rather than Southeast Asia, *B. barbigerum* provides a useful comparison at this point with *B. lobbii.* The flowers of *B. barbigerum* are undoubtedly one of the most bizarre structures in the plant world. This applies particularly to the dark brownish-red lip, which forms the largest part of the flower and which so greatly impressed Lindley. In a description published in the 'Botanical Register' in 1837, he considered the structure of the lip to be most extraordinary,

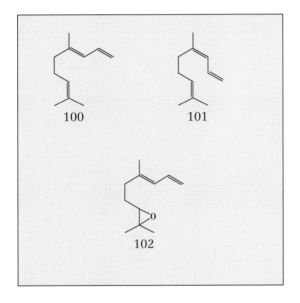

even for an orchid. He marvelled at its long, narrow, twisted body and the deep purple beard consisting of extremely fine, closely packed hairs. On the underside of the lip, some distance from the tip, he noted a second beard and, at the end of the lip, a 'brush' of long purple threads so unusually delicate that it moved with the slightest breath of air. Lindley described how these hairs—being of differing lengths and thicknesses—moved out of step with each other; indeed, he was struck by the very lively overall impression given by this movement together with that of the extremely mobile lip. Perhaps this explains why this gem among orchids used to be—and possibly still is —revered as a sacred plant in its homeland, where it is known as 'endua'.

Similar ciliated hairs also occur in typical 'fly flowers' of other families (cf. p.19, *Stapelia gettleffii*) and apparently contribute to the attractant effect produced by the shape, colour and scent of the relevant flower. If you come close enough to this fascinating mobile of *B. barbigerum,* you will notice a rather foul-smelling odour, reminiscent of overripe fruit and, slightly, of caproic acid.

Figures

157 *Bulbophyllum lobbii* Lindl.
158 *Bulbophyllum barbigerum* Lindl.

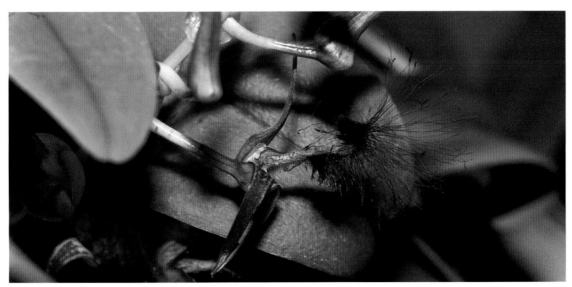

159

A very typical and beautiful representative of *Cirrhopetalum*, the genus that has been hived off from *Bulbophyllum*, is a species growing in a range extending from Thailand and Malaya to New Guinea, the Solomon Islands and Fiji,

Cirrhopetalum gracillium,

with its star-shaped flowers.

These flowers, with their unique shape and colouring so difficult to describe in words, emanate a rather restrained scent of algae and crustacea, a scent that can be experienced by the seashore at low tide. They are so fragile, however, that the trapping of analysable scent samples is a very difficult procedure, and the quantity of the analysis sample can be critical. Although certain olfactorily significant components were below the detection limit of the mass spectrometer, a major part of the scent is based on the identified compounds nonanal, (Z)-3-hexenol, (Z)-3-hexenyl tiglate, geranylacetone, acetic acid, isoleucine methyl ester and indole. Isoleucine methyl ester was recently identified by Joulain [50] in the scent of elder flowers *(Sambucus niger)*.

Also forming a semicircular umbel—an arrangement common to most of species of the genus—is the flower of

Cirrhopetalum robustum.

Orchids of the Indo-Australian tropics and subtropics

Already described on p. 33, this orchid grows in dense epiphytic vegetation on high trees in the rainforests of New Guinea. Its scent, or more accurately its stench, is one of the most penetrating produced by the orchid family (cf. p. 33). Primarily responsible for this olfactory experience, after which the nose requires a recovery period of at least 15 minutes, is a complex of acetic acid, butyric acid, 2-methylbutyric acid and valeric acid. Remarkably, however, aminic notes are also present, possibly deriving from N-methyl amine and N,N-dimethyl amine, although this could not be confirmed by analysis. Various amides that have been identified could, however, be taken as an indicator that, in this *Cirrhopetalum* species, amino acid catabolism also results in such free amines. These weak-smelling 'indicator compounds' include N,N-dimethyl acetamide (103), N-methyl acetamide (104), acetamide (105) and the corresponding formamides (cf. analysis, p. 207).

Of all the *Cirrhopetalum* species, which are invariably a delight to the eye,

Cirrhopetalum fascinor,

a native of Vietnam and Laos, is perhaps the most astonishing. In contrast to most of the other species in this genus, the raceme carries only one or two flowers, but these are of a remarkably strange design.

The upper sepal and the petals are decorated with red stripes and violet-purple

160

hairs. The lateral sepals are slightly pink, veined with purple lines and fused at their edges for much of their length, subsequently separating to form thin sepaline tails. These bizarre structures can reach up to 30 cm and more in length. An opening at their base allows the small – and once again highly mobile—lip to peep through. This 'seesaw' lip

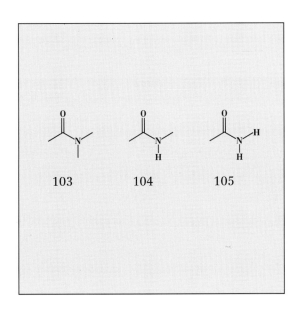

Figures
159 *Cirrhopetalum gracillium* (Rolfe) Rolfe
160 *Cirrhopetalum fascinor* Rolfe

comes into play once the flower's scent, which is reminiscent of stinkhorn and butyric acid, but also of fresh wood, has attracted the relevant fly species. Interestingly, the carrion-like aspects are concentrated in the lip area, whilst the quite separate and pleasantly woody note of caryophyllene epoxide tends to originate from the sepals (cf. analysis, p. 207). As described by Vogel [51], this phenomenon of two heterogeneous scent centres occurs to a much greater extent in *Cirrhopetalum ornatissimum* Rchb.f. According to Vogel, the 'tail osmophores' in this species possess a penetrating smell of codliver oil and fish remains, while the lip gives off an odour of trimethylamine.

Coelogyne Lindl.

The genus *Coelogyne* currently comprises about 200 mostly epiphytic species distributed right across the Indo-Australian floral region. Many of them are mountain-dwellers of the monsoon regions, and certain species of the southern Himalayas grow at altitudes as high as 3000 m. The invariably eyecatching flowers vary greatly in size and grow either singly or in clusters. Their scents, too, cover a very wide range, with 'spicy-floral' aspects much in evidence.

Thus, the Himalayan species *Coelogyne corymbosa* Lindl., growing at altitudes up to 3000 m, emits a very pleasant scent which is strongly reminiscent of Iris germanica. *Coelogyne cristata* Lindl., found on the southern slopes of the Nepalese Himalayas, smells of p-methyl anisole, anis aldehyde and vanilline. *Coelogyne fimbriata* Lindl. (China, Vietnam) presents a characteristic yeasty note, and is reminiscent of (Z)-4-decenal, whilst *Coelogyne flaccida* Lindl. calls to mind castoreum. Finally, the famous *Coelogyne pandurata* Lindl., known in Borneo as 'black orchid', emanates a scent that is reminiscent of vanilla and cinnamon.

I should like to describe

Coelogyne zurowetzii,

a native of the Mt. Kinabalu region (eastern

161

Malaysia), in rather more detail. It possesses a very unusual, though very pleasant, 'spicy-floral' scent with a leafy-green top note. It is based primarily on an accord composed of α-terpineol and methyl salicylate, and derives its special character from (Z)-3-hexenol, 3,5-dimethoxy toluene, eugenol, farnesol, methyl anthranilate and indole (cf. analysis, p. 209).

Cymbidium Sw.

The genus *Cymbidium* encompasses about 45 species that grow from India, across Southeast Asia to China and Japan, and also

162

as far south as Australia. Its main distribution area, however, is on the southern slopes of the Himalayan massif. Even more so than Cattleyas, hybrids of which have already become classic cut flowers, *Cymbidium* hybrids are now found in florist shops, offering flowers of an incredible variety and opulence. Without doubt, however, it is just as aesthetically pleasing to behold those natural forms which can be found—though unfortunately only rarely—in orchid collections. These generally grow epiphytically, but in some cases are terrestrial or lithophytic; many of them possess long, leathery leaves that sprout from the short, thickened 'pseudobulb-like' stems.

The 'scent connoisseur' will doubtless be especially attracted to the relatively small-flowered, but extremely elegant, Chinese and Japanese species, otherwise known as 'To-Yo-Ran' orchids. A specimen which could hardly be surpassed in elegance of scent and overall appearance is illustrated on p. 14. This is a much sought-after type of *Cymbidium goeringii* from Formosa, also known as *Cymbidium virescens*. Also presenting a fresh-floral scent that is strongly evocative of methyl jasmonate are *Cymbidium faberi* Rolfe and *Cymbidium kanran* Mak. Certain varieties of the latter species may also be characterized by 'ionone-floral' aspects. Finally, a pronounced 'ionone-floral' effect is also produced by the scent of *Cymbidium sinense* Willd., a species closely related to *C. kanran*. The various types of the so-called *Cymbidium harukanran*, a natural hybrid of *C. goeringii* and *C. kanran* [cf. 9], also belong in this second olfactory group of 'To-Yo-Ran' *Cymbidium* species.

The selection and cultivation, over many centuries, of these species of the *Cymbidium* genus in Japan and China have resulted in a huge variety of shapes and types. In their monograph on the *Cymbidium* genus [52], Du Puy and Cribb have taken this aspect into consideration and have attempted to classify the different varieties and synonyms.

I should like to present one particular representative of these highly fascinating orchids,

Cymbidium goeringii,

in rather greater detail. It grows terrestrially at altitudes of 500–3000 m in southern China, Formosa and Japan—here particularly in the coniferous forests close to the sea. The most sharply deviating variant can be found in Formosa (cf. fig. 2, p. 14), and has also been referred to as *C. formosanum* [cf. 52]. The

Figures
161 *Coelogyne zurowetzii* Carr.
162 *Cymbidium goeringii* (Rchb.f.) Rchb.f.

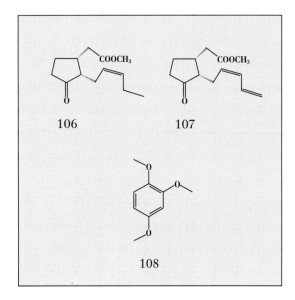

plant shown in Figure 162 is a common Chinese variety of *Cymbidium goeringii*. Its pleasant scent, reminiscent of lily of the valley, ripe lemons and a high amount of methyl jasmonate, is similar to that of the type found in Formosa, but is rather less elegant owing to a strange, almost imperceptible, herbaceous, aromatic facet.

As the analysis on p. 211 shows, this unique scent contains, together with methyl *cis*-(Z)-jasmonate (106), large quantities of a new jasmonoid compound with a structure corresponding to methyl *cis*-(Z)-dehydrojasmonate (107). This compound 107 is the main carrier of the scent of *C. goeringii*, and is principally responsible for the very pleasant, fresh-floral note, reminiscent of the scent of fully-ripe lemons; the main components nerolidol and (E,E)-farnesol provide the muguet-like contribution.

The previously mentioned herbaceous, aromatic aspect—which can easily be perceived as a rather negative facet in certain Japanese types of *C. goeringii*—is attributable to 1,2,4-trimethoxy benzene (108) and 1,2,3,5-tetramethoxy benzene.

Dendrobium Sw.

After *Bulbophyllum*, *Dendrobium* is probably the next largest orchid genus. The one thousand or so *Dendrobium* species are distributed almost throughout the tropical and subtropical belt of the Indo-Australian region which extends from southern Japan to New Zealand and includes numerous Pacific islands. Their preferred biotopes are the constantly humid, warm coastal regions plus the occasionally dry monsoon forests of the cooler mountain ranges. The species in this genus are extremely varied in terms of both plant structure and floral design. The shoots can be very small and succulent, or else can grow up to 2 m in length and be thick enough to form pseudobulbs. Other species are densely overgrown with fine hairs and create a very herbaceous impression. The size of the flowers can be equally variable, ranging from several millimetres up to 12 cm in diameter, and the colours pass from pure white via a delicate pink to a brilliant rich yellow and intense vermilion.

The few examples described below can give only a fleeting impression of this very diverse genus, although it should certainly suffice to call into question the opinion occasionally aired in the literature that the floral scents of this orchids are of minor significance. Of the 140 or so natural *Dendrobium* species evaluated, at least 40% were found to have moderately strong or strong scents, covering a range from decidedly fruity, via fruity-floral, 'ionone-floral', 'spicy-floral', woody-floral to herbaceous notes. Although very little has been written about the pollination of the *Dendrobium* orchids, we can conclude from the field observations cited by Pijl and Dodson [11] that most of the scented representatives are visited and pollinated by bee species. On the other hand, a tendency towards fly-pollination is apparent in certain small-flowered species—in the Australian *D. linguiforme*, for example—although these species do not display the corresponding syndrome.

The series of examples begins with a very remarkable species in all respects,

Dendrobium anosmum,

which was discovered in about 1840 near Manila, and described by Lindley in 1845 in the 'Botanical Register'. Its native territory

163

stretches from Laos and Vietnam, via the Malayan peninsula to New Guinea and the Philippines. The stems, reaching almost 2 m in length, tend to grow downwards and are subdivided by numerous nodes from which the magnificent purplish-lilac flowers emerge in pairs.

Just why Lindley should have chosen the name 'anosmum' in his original description remains a mystery. Perhaps he had been examining a scentless clone. Personally speaking, I have always found *D. anosmum* to possess a strong, very characteristic scent with, on the one hand, sweet and fruity-floral notes and, on the other, strongly herbaceous, almost rhubarb-like aspects. Chemically, too, the scent composition is very characteristic and unique amongst the orchids investigated. Almost 50% is accounted for by 2-pentadecanone, which is accompanied by a whole series of odd-numbered 2-alkanones (110) and certain 2-alkyl acetates derived from these.

A range of common floral scent constituents, including linalool, methyl benzoate, benzyl acetate, α-terpineol and indole, produces the 'aromatic-floral' part of the scent, whilst benzylacetone and dimethyl disulfide (109) are responsible for the exceptionally sweet top note (cf. analysis, p. 212). The widely-used synonym for *D. anosmum—D. superbum* Rchb.f.—certainly seems to be justified in every respect.

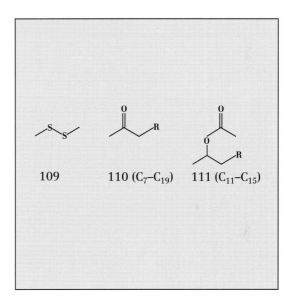

109   110 (C_7–C_{19})   111 (C_{11}–C_{15})

Various *Dendrobium* species found in extremely tropical habitats in New Guinea, the Pacific islands and the Sunda Islands are distinguished by their upright, curved petals which look rather like antelope horns. A typical representative of these 'antelope orchids' is an inhabitant of New Guinea and the surrounding islands,

Dendrobium antennatum.

The variety illustrated here, *d'albertsii,* develops up to 10 of these horned flowers,

Figure
163 *Dendrobium anosmum* Lindl.

which emanate a fresh, 'rosy-floral' scent based on linalool, geranial and geraniol and deriving its characteristic top note from a complex of esters, including butyl butyrate, butyl caproate and benzyl butyrate (cf. analysis, p. 212).

In stark contrast to the 'antelope orchids' is a group of small-flowered Australian *Dendrobium* species which, with their succulent growth form, have adapted to very extreme conditions and are able to withstand prolonged dry periods. A typical representative is

Dendrobium beckleri,

a species found in New South Wales and

Queensland. The flowers grow to about 2.5 cm in diameter and emit a very intense, somewhat aggressive scent, reminiscent of privet and honey. It is based on linalool, methyl phenylacetate, phenylacetaldehyde, methyl anisate and many other aromatic compounds, including vanilline (cf. analysis, p. 213).

A close relative of *D. beckleri,* inhabiting the same part of Australia, is

Dendrobium pugioniforme,

also known as the 'dagger orchid' on account of its sharp leaves. Its 'aromatic-floral' scent is strongly characterized by vanilla-like notes. In qualitative terms, it has a composition similar to that of *D. beckleri,* although quantitatively it is based on α-terpineol. A characteristic feature of this scent is the relatively large amount of phenylacetaldoxime—a compound occurring particularly in night-scented orchids—and the corresponding dehydration and oxidation product (cf. analysis, p. 216).

The flowers of

Dendrobium brymerianum,

a native of Laos, Burma and Thailand, conform rather more closely to the familiar image of this genus. Especially striking are the rich, orpiment-yellow colour of the flowers and the adornment of the lip with branched 'eyelashes' 1–2 cm in length. Although the 'aromatic-floral' scent of these flowers is highly characteristic, it is nevertheless composed of very familiar compounds with a primary accord consisting of decanal, methyl benzoate, benzyl acetate, methyl salicylate and indole. It is this, together with the main component linalool, which defines the character of the scent (cf. analysis, p. 213).

One group of interestingly scented *Dendrobium* species is characterized, in the non-flowering state, by a fine covering of dense black hair on the young shoots. The first representative of these species, which belong to the *Nigrohirsutae* section, is

Dendrobium carniferum.

This orchid is found in Assam, Laos, northern Burma and northern Thailand at altitudes of 700–1000 m. The flowers measure

165

166

5–6 cm across and are particularly notable for their contrasting colours (white-to-cream and orange-red) and their scent, which, apart from the linalool-like floral basic accord, is strongly reminiscent of the ester part of the orange flavour.

Ethyl propionate, ethyl butyrate and ethyl crotonate are mainly responsible for this typical aspect (cf. analysis, p. 213).

The same distribution area is also home to the closely related, and also black-haired, species *Dendrobium draconis* Rchb.f. Its leaves are coarser, however, and the pure white flowers, displaying vermilion markings in the throat of the lip, not only appear at the end of the stems but also sprout from the nodes and extend over the full length of the thick, pseudobulb-like stems. Their scent is related to that of *D. carniferum,* but is more evocative of mandarines and tangerines. It probably contains fairly large quantities of the typical 'citrus aldehydes'. Accordingly, it forms a neat transition to the scent of the closely related

Dendrobium williamsonii.

Figures
164 *Dendrobium antennatum* var. *d'albertsii* Lindl.
165 *Dendrobium beckleri* F. Mueller
166 *Dendrobium pugioniforme* Cunn.
167 *Dendrobium brymerianum* Rchb.f.

167

168

Although this scent also contains fairly large amounts of linalool, in olfactory respects it bears the strong imprint of a series of aldehydes, with dodecanal and tetradecanal playing a particularly important role. These typical citrus aspects are accompanied by spicy notes, lending the scent a rather unusual, but nonetheless appealing, character (cf. analysis, p. 217).

D. williamsonii grows on the southern slopes of the Himalayas at slightly higher altitudes than the preceding black-haired *Dendrobium* species. Whilst its flowers possess a similar colouring scheme, they can easily be distinguished from the three other species mentioned here by the accentuated 'eyelashes' along the lip.

Another member of this group is *Dendrobium margaritaceum* Finet., an inhabitant of the monsoon forests of Thailand. The white flowers grow to about 3 cm in diameter and the throat of the lip is a yellowish-orange. They emit a full-bodied 'rosy-floral' scent which is based on geraniol but—owing to the high citral and methyl 2-methylbutyrate content—is also reminiscent of aspects of lemon and orange. This delightful orchid is sometimes confused with the closely related *Dendrobium bellatulum* Rolfe, which grows in the same area. Although its scent is very similar, it does contain an additional yeast-like note.

169

While these representatives of the genus bear only small numbers of flowers, the pendulous racemes of

Dendrobium chrysotoxum

bear flowers in abundance. The golden-yellow flowers, with an orange tinge towards the throat of the lip, reach a diameter of 4–5 cm, and are unmistakable in appearance, with their fine hair-covering and finely-fringed lip margins. This splendid orchid grows on high

Figures
168 *Dendrobium carniferum* Rchb.f.
169 *Dendrobium williamsonii* Day et Rchb.f.
170 *Dendrobium chrysotoxum* Lindl.

170

trees in the monsoon mountain forests of Burma, Thailand, central Laos and the southern Chinese province of Yünnan. Its rather heterogeneous scent is initially reminiscent of melons, subsequently presenting jasmine-like and spicy notes. The main components have been identified as α-pinene, myrcene, hexyl acetate and geranyl acetate. Notable minor components of olfactory significance include verbenone, ethyl cinnamate, eugenol, phenylethyl tiglate, dihydroactinidiolide and methyl cis-(Z)-jasmonate (cf. analysis, p. 214).

A number of other scented yellow-flowering species that are more or less closely related to *D. chrysotoxum* inhabit the area between the southern Himalayas and north Thailand and Burma. The scents of several representatives are briefly summarized below.

Dendrobium chrysanthum Lindl.:
pronounced 'aromatic-floral'; melon aspect in the top note similar to that of *D. chrysotoxum*.

Dendrobium densiflorum Lindl.:
basic floral accord based on linalool and characterized by honey-like and leaf-green to aldehydic notes.

Dendrobium griffithianum Lindl.:
pronounced 'aromatic spicy-floral', strongly reminiscent of eugenol and benzyl acetate.

Dendrobium fimbriatum Hook.:
similar to *D. griffithianum*, definite eugenol and vanilline note.

And finally, one of the most beautiful species in the genus, the deep canary-yellow

Dendrobium harveyanum Rchb.:
highly multifaceted, reminiscent of honey, primula, cloves and (slightly) of mimosa.

One of the strangest scents in this genus is given off by

Dendrobium delacourii,

a species native to Burma and Thailand.

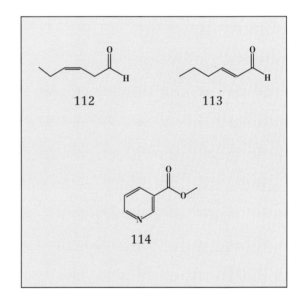

It produces a very herbaceous and green effect, strongly redolent of rhubarb, and is based on large quantities of (Z)-3-hexenal (112), (E)-2-hexenal (113) and the corresponding alcohols and benzyl acetates. Methyl nicotate is a highly characteristic compound in this scent, being present in relatively large amounts (cf. analysis, p. 214).

Dendrobium lichenastrum

belongs to the same section as the two previously mentioned species *D. beckleri* (p. 154) and *D. pugioniforme* (p. 154) and, like these, also resides in tropical eastern Australia. Possibly the most peculiar *Dendrobium*, this species has no pseudobulb, possessing instead very succulent, tightly packed leaves just 0.5–1 cm in length which sprout directly from a very thin, branched rhizome. Emerging individually from this wickerwork formation are the very small, translucent flowers, which grow to just 4 to 5 mm in diameter and are white to pink in colour. Despite their small stature and modest appearance, they do not merely manage to attract the rapt attention of the relevant pollinators. The human observer, too, feels sure that he has often experienced a very similar smell in the most diverse regions of the world; or was it perhaps a very similar taste, in the form of a well-known drink? In the case of *D. lichenastrum*, this olfactory impression is based on an accord consisting of limonene, linalool and geranyl acetate, deriving its character from the hesperidic notes of citral, decanal and dodecanal and the aromatic-spicy notes of cinnamic aldehyde, methyl cinnamate and vanilline (cf. analysis, p. 215).

Another member of the same group is *D. rigidum* R.Br., which grows in northern Queensland and New Guinea, particularly on mangroves. The floral scent of this species is strongly reminiscent of coumarin and anis aldehyde.

Also at home in tropical eastern Australia is

Dendrobium monophyllum,

which possesses one of the most delightful of all orchid scents. It, too, is a rather small orchid: the pseudobulbs are 1–3 cm long and each bear one leaf 5–10 cm in length. The forward-leaning inflorescence emerges from the leaf sheaths and presents up to 15 flowers, each about 2 cm in width. Particularly around midday, these emit a very pleasant and interesting scent which has 'rosy-floral', 'aromatic ionone-floral' and fruity-herbaceous facets. The characteristic basic accord of this scent is formed by linalool, methyl citronellate (methyl 2,7-dimethyl-6-octenoate) and β-ionone and its derivatives, rounded off by other compounds, including methyl cinnamate and indole (cf. analysis, p. 216).

The main component of the scent of *Dendrobium monophyllum*, methyl citronellate, is much less widespread in nature than its precursor methyl geranate. In the orchid scents investigated to date, it has only been additionally identified as a minor component in *Encyclia citrina*.

A native of Japan, Korea and Formosa,

Dendrobium moniliforme

is one of the orchids described by Carolus Linnaeus himself. Belonging to the 'To-Yo-Ran' orchids, it had already been cultivated in Japan for many centuries previously, and numerous varieties can now be admired in collections of these types of orchids.

The flowers, growing to 3–4.5 cm in diameter, are normally coloured from snow-white to varying shades of pink. Their scents can vary greatly, depending on the variety or clone. All, however, possess a pleasant basic accord composed mainly of linalool, phenylethyl alcohol and farnesol but with specific characteristics which—depending on the plant—may, for example, derive from the aromatic, warm-herbaceous note of hydroquinone dimethyl ether or from methyl anthranilate, which is reminiscent of Concord grapes. A comparison of the analytical results

Figure
171 *Dendrobium delacourii* Guill.

Orchids of the Indo-Australian tropics and subtropics

172

Orchids of the Indo-Australian tropics and subtropics

173

174

for a pure white and a pink type is shown on p. 215.

Constrasting starkly with this 'To-Yo-Ran' orchid in both visual and olfactory terms is

Dendrobium trigonopus,

an inhabitant of Burma, Laos, Thailand and southeast China. The golden-yellow flowers with their greenish interiors are pervaded by a very characteristic scent, arising from the interaction between herbaceous-woody and aromatic-floral complexes.

The herbaceous-woody part of this scent consists primarily of *cis*-verbenol (115), verbenone (116), *trans*-verbenone epoxide (117) and a further unidentified monoterpene

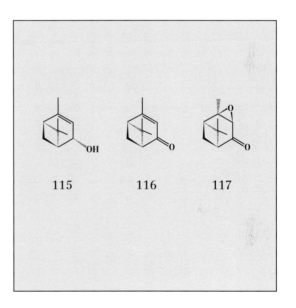

epoxy ketone, whilst the aromatic-floral part is largely based on benzaldehyde, methyl benzoate, phenylacetaldehyde and geraniol (cf. analysis, p. 216). *cis*-Verbenol (115) also occurs in smaller quantities in the scent of *Huntleya meleagris* (p. 56), which is dominated by the corresponding *trans*-isomer,

Figures
172 *Dendrobium lichenastrum* (F. Mueller) Kraenzl.
173 *Dendrobium monophyllum* F. Mueller
174 *Dendrobium moniliforme* (L.) Sw.

Orchids of the Indo-Australian tropics and subtropics

whilst *trans*-verbenone epoxide (117) has not been identified in any other orchid scent.

A species occurring in northern and northeast Thailand and the bordering provinces of Laos,

Dendrobium unicum,

should dispel any doubts about the wealth of scents occurring within this genus.

This small epiphyte, with leaves up to 5 cm across, was first discovered in the 1960s. It lives in maquis-like mountain forests and is closely related to *Dendrobium arachnites* Rchb.f., a species growing mainly in Burma. With its retroflexed sepals and petals and non-resupinated lip, the flowers of *D. unicum* bear a striking resemblance to those of a coral tree species (*Erythina* sp.), thus making this orchid even more of a sensational find.

The species name *unicum* could also refer to the scent, which perfectly matches the brilliant orange colour of the flowers. Its basic concept derives from the interaction of orange and apricot to peach-like notes, and is finally complemented by 'aromatic-floral' aspects. The main component of this unmistakable scent, (E)-3-methyl-4-decen-1-ol (118), is also unique, and is accompanied by smaller quantities of the corresponding aldehyde (119).

These two compounds, which have not yet been identified in any other orchid, form the

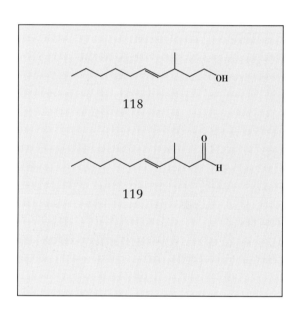

characteristic fruity-floral basic scent. This in turn is complemented by the apricot and peach notes of γ-decalactone and γ-dodecalactone and by the cinnamic to aromatic-floral notes of cinnamic aldehyde, methyl cinnamate and 2-amino benzaldehyde (cf. analysis, p.217). In contrast to 119, (E)-3-methyl-4-decen-1-ol (118) has been mentioned as an ingredient of nutmeg oil [53], the Likens-Nickerson extract of the flowers of

Figures
175 *Dendrobium trigonopus* Rchb.f.
176 *Dendrobium unicum* Seidenf.

Orchids of the Indo-Australian tropics and subtropics

Acacia farnesiana [54] and the flower concrete of *Michelia champaca* [55].

An olfactory treat of a completely different sort is provided by the last representative of this genus illustrated here,

Dendrobium virgineum.

At around midday, the large, aesthetically pleasing, ivory-white flowers emanate a very gentle, velvety-woody scent which is discreetly accompanied by aldehydic and 'white-floral' to 'rosy-floral' notes. This pleasant woody note, produced largely by caryophyllene epoxide and humulene epoxide II, increases in intensity until the early evening, eventually dominating the scent completely (cf. analysis, p. 217).

D. virgineum grows in northwest Malaya and Burma, and seems to be fairly variable. For example, one type with glossier petals and sepals and more richly contrasting lip colouring produces a much more floral scent that is particularly evocative of white peonies.

Dendrochilum Bl.

The genus *Dendrochilum* encompasses about 120 species distributed throughout Southeast Asia, with the greatest concentration of species occurring in the mountains of Sumatra and Borneo. Many of these small to medium-sized epiphytes grow endemically in poorly accessible regions and, consequently, are rarely encountered in collections. Their flower-bearing stalks range from a few centimetres to 50 cm in length, depending on the species, and sprout laterally from the pseudobulbs. They are densely covered with small, delicate flowers arranged in two rows.

178a

177

One of the better-known, strongly scented species,

Dendrochilum cobbianum,

grows on the island of Luzon. This orchid can develop numerous 'flower chains' up to 50 cm long which are translucent in sunlight and

Orchids of the Indo-Australian tropics and subtropics

178b

179 Photo H.P. Schumacher

envelop the plant in a curtain of flowers and scent. It is a very fresh scent based on the transparent, floral-terpenic note of (E)-ocimene and various derivatives, and deriving its special character from a large number of smaller components, including phenylacetaldehyde, citral, β-ionone, 2-amino benzaldehyde and vanilline (cf. analysis, p. 218).

Another species frequently encountered in collections, *Dendrochilum glumaceum* Lindl., has a similar distribution area. Its scent is extraordinarily sweet and aromatic-floral, and is strongly pervaded by anis aldehyde, geranyl acetate and veratrol (1,2-dimethoxy benzene).

Epigeneium Gagn.

This small genus is very closely related to *Dendrobium* and encompasses about 35 rare epiphytic and lithophytic species that are of considerable interest. It inhabits a range extending from southern China and India via the Philippines to New Guinea. A particularly well-known representative is a native of the mountains of Luzon,

Epigeneium lyonii.

Figures
177 *Dendrobium virgineum* Rchb.f.
178a+b *Dendrochilum cobbianum* Rchb.f.
179 *Epigeneium lyonii* (Ames) Summerh.

The overhanging inflorescence, up to 50 cm in length, bears 5–20 individual flowers, each growing up to 8 cm in diameter. The scent they emit is probably unique among flowering plants. It is completely dominated by γ-octalactone and, consequently, is reminiscent of coconut, caramel and tonka beans (cf. analysis, p. 223).

Eria Lindl.

With over 500 epiphytic or terrestrial species covering a range from India to Fiji, *Eria* is one of the major genera of the orchid family. Many *Eria* species possess small flowers that are rather unremarkable in shape and colouring; their scents often show a floral aspect with herbaceous to green notes reminiscent of yeast, algae or sperm. Several species, however, possess eyecatching medium-sized flowers and, being rare representatives of the genus, have found their way into orchid collections. Perhaps the most attractive of these is

Eria hyacinthoides,

developing inflorescences up to 30 cm long which, like hyacinths, are densely covered with 30 to 40 white flowers.

Their aromatic-floral and herbaceous scent is based on linalool and hydroquinone dimethyl ether, displays a yeast note produced by (Z)-4-decenal and related compounds, and is highly diffusive, thanks to relatively large quantities of dihydro-β-ionone (cf. analysis, p. 224).

We have not yet been able to establish the compound responsible for the sperm-like note that was also detected.

This note is virtually dominant in

Liparis viridiflora,

a species occurring virtually throughout the Indo-Australian region. Its strange scent is also strongly influenced by the aldehydes (E)-2-nonenal and (E,Z)-2,6-nonadienal (120), which are reminiscent of cucumber and violet leaves. Other noteworthy constituents are the two diastereoisomers of chalcogran (121a/b, cf. analysis, p. 228). These spiroketals 121a/b are known to be biologically active compounds of the aggregation pheromone of the bark beetle *Pityogenes chalcographus* [40].

Interestingly, the scent of *Liparis viridiflora* contains surprisingly large quantities of two other isomeric compounds that have been mentioned elsewhere as insect pheromones. These are the (E)- and (Z)-isomers of 7-methyl-1,6-dioxaspiro[4.5]decane, which have been identified by Franke and co-authors [40] in the pentane extract of female workers of *Paravespula vulgaris* L.

We have found the two chalcograns (121a/b) in roughly equal amounts in the floral scents of *Eria hyacinthoides, Oncidium ornitho-*

180

Orchids of the Indo-Australian tropics and subtropics

181

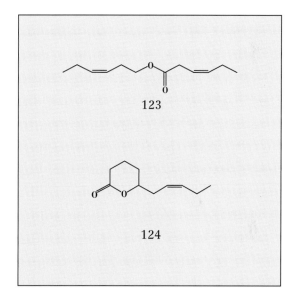

Neofinetia H.H.Hu

This genus consists of just one species, which is also the only east Asian 'moth orchid' and may be likened to the African angraecoid orchids: the extremely delicate and very pleasantly night-scented

Neofinetia falcata.

This little gem resides in China, Korea, the Ryukyu Islands and, especially, in Japan, where it has been cultivated for centuries as one of the 'To-Yo-Ran' orchids. It was named in honour of the French botanist Achille Finet (1862–1913), who specialized in the study of Japanese and Chinese orchids.

rhynchum (p. 118) and, together with the spiroketals 122a/b, in that of *Dracaena fragrans* [16].

The 250 or so species of the *Liparis* genus are distributed across all the orchid vegetation zones, but are commonest in the Indo-Australian region. Their flowers are mostly small to minute and are predominantly greenish, white or cream-coloured. Their scents cover a spectrum similar to that of the *Eria* species, and are frequently reminiscent of algae, yeast, crustacea and similar odours that are probably highly attractive to fly and midge species, but which certainly also hold many surprises in store for the chemist.

The pure-white flowers, reaching 2.5–3 cm in diameter, bear an elegantly curved spur, some 4–6 cm in length, and—particularly during the night—give off an intense and extraordinarily pleasant 'white-floral' scent, reminiscent of aspects of honeysuckle, tuberose and gardenia. Its main components are linalool, methyl benzoate, benzyl alcohol, (Z)-3-hexenyl tiglate and benzyl tiglate, which, however, are complemented by a

Figures
180 *Eria hyacinthoides* (Bl.) Lindl.
181 *Liparis viridiflora* (Bl.) Lindl.

167

wide range of olfactorily important compounds.

These include tiglyl tiglate, a substance also found in the scent of *Plectrelminthus caudatus* (94, p. 138), (Z)-3-hexenyl (Z)-3-hexenoate (123)—with its characteristic green note reminiscent of bamboo shoots—and jasmine lactone (124), together with other C_{10}-lactones, indole and relatively large quantities of vanilline (cf. analysis, p. 235).

Phalaenopsis Bl.

The 40 or so epiphytic species of the *Phalaenopsis* genus are typical inhabitants of the hot and humid rainforests of Southeast Asia and northern Australia. Instead of pseudobulbs, they possess a very short stem with large leathery leaves. Their flowers are in most cases splendidly coloured, though white predominates in several species. The pure white, large-flowered *Phalaenopsis amabilis* has often been crossed with other species and is now available as a cut flower hybrid at almost every florist's. However, the incomparable elegance of the natural form, several subspecies of which grow in Indonesia, New Guinea and tropical northern Australia, is scarcely bettered by any of these hybrids. Personally, I have only ever encountered scentless examples of this *Phalaenopsis*,

otherwise known as the 'moon orchid'. According to Pijl and Dodson [11] however, a variety growing in New Guinea emits a strong, sweet-floral scent. 'Aromatic-floral' to 'spicy-floral' scents are particularly common in this genus, for example in *P. fasciata* Rchb.f. and *P. hieroglyphica* (Rchb.f.) Sweet, as are 'rosy-floral' scents, e.g. in *P. schilleriana* Rchb.f.

The latter scent group includes

Phalaenopsis violacea,

which will now be described in more detail. It grows in Sumatra, Borneo and the Malayan peninsula in shady locations close to rivers, and occurs both as a 'Malayan type' (Figure 183) and a 'Borneo type' (Figure 184).

Apart from minor morphological differences in floral structure, which are of secondary significance for many taxonomists, these two types differ primarily in floral colouring and scent. Whilst the 'Borneo type' emits a pure 'rosy-floral' scent based on linalool, citronellol, nerol and geraniol (cf. analysis, p. 239), the 'Malayan type' displays an additional, harmoniously inte-

Figure
182 *Neofinetia falcata* (Thunb.) H.H.Hu
183 *Phalaenopsis violacea* Witte, Malaya type
184 *Phalaenopsis violacea* Witte, Borneo type

Orchids of the Indo-Australian tropics and subtropics

185

grated cinnamic note which can be traced to considerable quantities of cinnamyl acetate, cinnamic alcohol and cinnamic aldehyde (cf. analysis, p. 240).

In chemical terms, the scent is also distinguished by the main component, elemicine (71). This compound is also contained in large quantities in the scents of *Miltoniopsis phalaenopsis* (p. 109) and *Laelia perinii* (p. 95).

Rhynchostylis Bl.

The genus *Rhynchostylis* includes just four species that grow in Thailand, Malaya, Indonesia and the Philippines. Their general appearance is very similar to that of *Aerides*, although *Rhynchostylis* species possess shorter stems and thicker leaves, and differ with regard to the structure of the column and spur.

Extremely attractive in every respect is the Thai species

Rhynchostylis coelestis.

In contrast to the three other species, this orchid develops upright, rather than pendulous, inflorescences. The waxy, indigo-blue and white flowers emanate a homogeneous and very full-bodied, though not aggressive, scent which—perhaps more than any other species—arouses the association of 'damp petals'. Its overall character could be described as fresh-floral, and it is based on very large amounts of farnesal, farnesol and ocimene, rounded off by smaller amounts of compounds such as linalool, α-terpineol, benzyl acetate and anethole (cf. analysis, p. 243).

Trichoglottis Bl.

Of the 60 or so species in this genus scattered throughout Southeast Asia and Polynesia, only a few representatives are found in collections. One of these is the relatively common species

Trichoglottis philippinensis,

186

which grows at low altitudes on the Philippine islands of Luzon, Mindanao and Negros, and forms long climbing creepers. Its herbaceous-floral scent contains α-farnesene and linalool as the main components and is especially characterized by p-methyl anisole, benzaldehyde, 3-phenylpropanal and 3-phenylpropanol (cf. analysis, p. 245). Similar scents are also encountered in the neotropical *Brassia* species (p. 60).

Figures
185 *Rhynchostylis coelestis* Rchb.f.
186 *Trichoglottis philippinensis* Lindl.

187　　　　　　　　　　　　　　　　　　　　　　　　　　　　　　Photo Ch. Weymuth

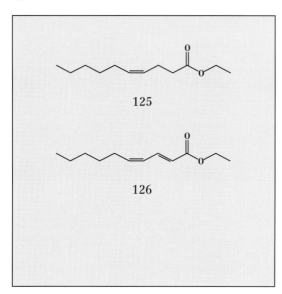

Trixspermum Bl.

Trixspermum is another genus rarely seen in collections. It covers about 80 epiphytic, generally small-flowered species that grow from Thailand to Indonesia and are particularly numerous on Sumatra.

Trixspermum arachnites (Bl.) Rchb.f.

occurs throughout the distribution area of the genus and, since it measures up to 5 cm across, is counted as one of the large-flowered representatives. Its intensive, very characteristic scent exhibits green-herbaceous and fruity aspects in addition to floral notes that are rather reminiscent of gardenia.

This scent is based on the main components *cis*-linalool oxide and nerolidol, and derives its character mainly from (E,Z)-2,4-decadienal, ethyl (Z)-4-decenoate (125), ethyl (E,Z)-2,4-decadienoate (126) and the corresponding (E,E)-isomers, plus γ-decalactone (cf. analysis, p. 245).

Whilst some of the investigated orchid scents contain the methyl esters of (Z)-4-decenoic acid and (E,Z)-2,4-decadienoic acid as minor components, we have identified the corresponding ethyl esters 125 and 126 only in the scent of *Trixspermum arachnites*. Ethyl (E,Z)-2,4-decadienoate (126) and the corresponding methyl ester are responsible for the typical aroma of Bartlett pears; the ethyl ester 126, in particular, is employed in the flavourings and perfumery industries under the name 'pear ester'.

Vanda Jones

The genus *Vanda* includes about 70 species which inhabit an area stretching from the humid, hot biotopes of the Sunda Islands to the cool heights of the Himalayan massif. The appearance of these epiphytic and lithophytic species ranges from that of miniature orchids to profusely proliferating giant plants. Like the angraecoid orchids (p. 123), however, all species exhibit a monopodial form of growth. The name 'Vanda' is a Sanskrit word applied to the species *Vanda tessellata* described below.

One of the small representatives of the genus is the graceful

Vanda coerulescens,

which grows in Burma and Thailand at altitudes of 300–500 m. The whole plant rarely exeeds 20 cm in height, and develops inflorescences of approximately the same length covered with numerous carmine-violet blue flowers of about 3 cm in diameter. These are often scentless, but certain clones emit the characteristic scent of the Concord grape. As one might expect, methyl anthranilate is the dominant component, both in quantitative

189 Photo Ch. Weymuth

and qualitative respects. The scent of *Vanda coerulescens* also contains roughly the same quantity of a methyl decatrienoate whose structure has yet to be established, accompanied by both isomers of methyl 2,4-decadienoate, which have already been encountered in several orchid scents (cf. analysis, p. 245). In addition, this unusual scent contains relatively large amounts of 2-amino benzaldehyde, (Z)-3-hexenal and (Z)-3-

Figures
187 *Trixspermum arachnites* (Bl.) Rchb.f.
188 *Vanda coerulescens* Griff.
189 *Vanda denisoniana* (Bens.) Rchb.f.
190 *Vanda tessellata* (Roxb.) G. Don (p. 174)

Orchids of the Indo-Australian tropics and subtropics

190

hexenol, which are also olfactorily important components.

A much larger representative of the genus,

Vanda denisoniana,

grows in the Arracan mountains of Burma. Its flowers can vary in colour from greenish-white via pure white to ivory yellow. The very appealing scent of the type illustrated is based on linalool and ocimene, and is chiefly characterized by 2-amino benzaldehyde and methyl anthranilate (cf. analysis, p. 245).

Encountered with greater frequency in collections is an inhabitant of the same region, *Vanda denisoniana* var. *hebraica,* whose yellow-coloured sepals and petals exhibit markings that closely resemble Hebrew lettering. Its very intense 'rosy-floral' scent differs greatly in quantitative and olfactory respects from that of the normal form, and is based largely on nerol, geraniol and myrcene.

Significantly more variable in the colouring and shape of its flowers is the species

Vanda tessellata,

an inhabitant of Sri Lanka, India and Burma. This strongly-scented species covers a particularly large number of varieties and is especially common in the dry zone of Sri Lanka. Remarkably, this epiphytic orchid appears to be connected in some way with human habitation, since it rarely grows far from settlements and is never found in dense jungle [56]. The example illustrated is the most typical form and has grey-green-brown coloured flowers some 5 cm in diameter and a deep lavendar-coloured lip. Its very intense aromatic-floral scent is based on linalool, methyl benzoate, cinnamic aldehyde and methyl cinnamate, deriving its ultimate character from a variety of smaller components, including benzyl acetate, α-ionone, 3-phenylpropanal, p-cresol and indole (cf. analysis, p. 246).

The scents of two other types of *V. tessallata* are based, in one case, on (E)-ocimene, methyl salicylate and cinnamic aldehyde and, in the other, on (E)-ocimene, methyl benzoate, benzyl alcohol, phenylethyl alcohol and the benzoates of the last two compounds. All possible combinations of these three 'scent skeletons' seem to occur within this species.

Some European orchids

After travelling through the principal vegetation zones of the tropics and subtropics, we may have become so 'spoilt' by so many large and magnificent species that we can no longer really appreciate the terrestrial species of Europe, most of which bear small flowers. However, closer examination of these flowers opens up an equally fascinating world of colours and shapes, and reveals that their scents also cover an amazingly broad spectrum. To keep matters in proportion, I should now like to present just a few of the well-known representatives of the 70–80 central European species.

Anyone who was brought up in Switzerland would almost certainly start with the most popular floral resident of the Alps,

Nigritella nigra,

commonly known here as 'Männertreu' or 'Orchis vanillé'. Just as perfumery could not manage without the scents of rose and jasmine, summer in the mountains—with all its contrasts and teeming life—would not be complete without the visual and olfactory impressions created by this delightful orchid. *Nigritella nigra* grows mainly in sunny, dry alpine meadows at altitudes of 1000 to 2800 m (cf. Figure 14, p. 23, and Figure 30, p. 39).

The hemispherical to spherical inflorescences consist of 20–70 small, non-resupinated, generally dark purple flowers. Very rarely encountered are not only albinotic types, but also pink, orange and yellow coloured variants. The scent is particularly fascinating, the usual form producing an 'aromatic spicy-floral' effect which is strongly reminiscent of vanilla and cocoa. As the scent analysis described on page 235 shows, this scent is based on phenylethyl alcohol and benzyl alcohol and does in fact also contain a series of compounds also identified in vanilla and cocoa extracts as vanilline, vanillyl ethyl ether, phenylacetaldehyde, isovaleraldehyde and benzyl isovalerate, plus further valerates.

The natural habitat of *Nigritella nigra* is also home to another, relatively common, terrestrial orchid with a typical 'aromatic spicy-floral' scent: the species *Gymnadenia conopea,* discussed below. Occasionally, these two species, belonging to different genera, are combined to form a genus hybrid known as *Gymnigritella suaveolens*. The flowers of this hybrid, which are usually of a brilliant carmine colour, reflect the morphological and olfactory features of the two parents in an almost idealized manner.

Gymnadenia conopea

is a relatively common plant in Europe (except in the southern regions) and is also found across the temperate zones of Eurasia and Japan. It is one of the most variable of the terrestrial orchids in this huge distribution area, and has adapted itself to the most diverse biotopes, from lowlands to alpine habitats. Its stature can vary from compact to lanky, and the 5–30 cm long inflorescence carries numerous flowers which are up to 1.5 cm across and cover the full range of hues from intense reddish-lilac to light pink. *Gymnadenia conopea* varies just as greatly in respect of its floral scent.

As already described in somewhat greater detail on page 27, the scent composition can vary considerably even within a 'homogeneous' population. As a result, the 'aromatic spicy-floral' basic scent in the specimens investigated can be characterized, in one group of plants, by eugenol and cinnamic alcohol and, in another, by benzyl acetate and related compounds. Moreover, low-lying biotopes can be home to large, light pinkish-lilac coloured types whose scent is strongly reminiscent of the waxy and slightly metallic aspects of capric acid and lauric acid, whilst the scent of certain small alpine

Some European orchids

191

olfactorily important components such as citral, cinnamic aldehyde, cinnamic alcohol, eugenol and vanilline. In chemical terms, the scent of *Gymnadenia odoratissima* is quite closely related to that of *Nigritella nigra*.

Two typical moth orchids

The comparatively orchid-deficient region of central Europe is in fact home to two typical moth orchids, namely *Platanthera bifolia* and *Platanthera chlorantha,* whose flowers emanate intense and appealing scents during the night. Just like the angraecoid moth orchids of Africa (p. 123), their flowers are white in colour and bear a long spur. These two species are visited especially frequently by relatively large moths, e.g. the euphorbia sphinx (*Hyles euphorbiae* L.), the vine hawk-moth (*Deilephilia elpenor* L.) and the pine hawk-moth (*Hyloicus pinastri* L.). The genus *Platanthera* encompasses about 80 species distributed throughout the northern hemisphere, and these vary greatly in overall appearance.

Of the two European species to be presented in some detail,

Platanthera bifolia

is the more common. Like *Gymnadenia conopea,* it has adapted itself to a variety of biotopes, from lowlands to alpine regions. Depending on its location, this eyecatching orchid (known under numerous popular names) flowers between mid-May and the beginning of August, and the opportunity should not be missed to view it at least once during this period either at, or just after, dusk. Owing to the white colour of the flower and its appealing scent—both discernible from afar—mean that this graceful orchid is located fairly easily. In Scandinavia, *Platanthera bifolia* flowers at the time of the midsummer festival. Given the name 'Nattviole' (night

types has a definite vanilla note (cf. analysis, p. 225).

The latter aspect is fairly pronounced in the scent of the second European species of this genus, the smaller-flowered *Gymnadenia odoratissima*. As its name suggests, this much rarer species, which grows only in Europe, emits a very pleasant scent that could also be defined as 'aromatic spicy-floral'. By and large, it contains the same compounds found in *Gymnadenia conopea* though with significant quantitative differences. It is based on a basic accord consisting of the main components phenylethyl alcohol, benzyl alcohol and benzaldehyde and deriving its pleasant character from a series of minor but

Figure
191 *Gymnadenia conopea* (L.) R.Br.

violet), it is seen as symbolizing the joys of summer.

Although the German popular name 'Waldhyazinthe' (wood hyacinth) and many literature sources ascribe a hyacinth-like scent to this orchid, it is more reminiscent of certain lily and honeysuckle species and, in very general terms, possesses a basic scent concept of the 'white-floral' type. In the case of the type illustrated, this is attributable to linalool, methyl benzoate and methyl salicylate. Here again, however, many compounds present in fairly small quantities give the scent its particular character. Special significance is ascribed to the four isomers of lilac aldehyde (127 a–d), the corresponding alcohols and their acetates, cinnamic aldehyde, geraniol, eugenol and vanilline (cf. analysis, p. 240).

But like *Gymnadenia conopea*, for example, *Platanthera bifolia* is variable not just in its general appearance but also in its scent composition. Thus, Nilsson [57] investigated a Swedish type that contained neither the isomers of lilac aldehyde (127 a–d) nor those of the corresponding alcohol, whilst Tollsten and Bergström [58], in a systematic investigation of various Swedish populations, observed that the total content of these compounds can vary from virtually 0% to 40%, depending on the plant. Wakayama and co-authors [59–60] discovered lilac alcohols and lilac aldehydes in the floral extract of the lilac (*Syringa vulgaris* L.), and elucidated their structure, at the beginning of the 1970s. The lilac aldehydes a–d (127 a–d) are listed according to their retention times on medium to polar GC columns.

No less variable in its scent composition is

Platanthera chlorantha,

which differs from *P. bifolia* mainly on account of its widely splayed pollinia and the slightly different shape of its spur. For some inexplicable reason, *P. chlorantha* is often described in the literature as being scentless or weakly-scented. In fact, it possesses, to an even greater extent, the characteristic feature of the angraecoid orchids—namely, that of being scentless during the day but giving off an intense 'white-floral' scent at night. In the type illustrated here, this scent is based on linalool and methyl benzoate, accompanied by a velvety-rosy note which— as in the neotropical species *Brassavola nodosa* (p. 58)—is derived from a high geraniol content. Interestingly, Tollsten and Bergström [58] found clones of *Platanthera chlorantha* on Öland (Sweden) whose floral scent contained up to 90% lilac aldehyde (127 a–d), including its derivatives. By contrast, in other populations on the Swedish mainland, they found plants with practically the same scent composition as the type discussed here (cf. analysis, p. 241).

127 a 127 b 127 c 127 d

Some European orchids

192

193

The very last species to be presented in this section,

Himantoglossum hircinum,

(lizard orchid) was probably the first orchid which, in my youth, I strongly associated with the concept of 'scent'. It was an unusual scent, however, and one which aroused my curiosity and perhaps puzzled me a little as well. Those who are ignorant of this stately orchid—which has an affinity for warm, sunny, grassy slopes —could easily overlook it, despite the fact that it stands up to 1 metre tall. This is due to the pale green to olive or brownish-white colouring of its individual flowers, up to 80 of which can be crowded onto the inflorescence, itself often 30 cm in length. But once the wind blows its scent in your direction, it will no longer escape your notice. It seems unlikely that such an exotic-looking plant should feel at home in these latitudes; particularly puzzling are the numerous lips, which are up to 7 cm long and turned in a screw-like fashion. But then the spectator becomes utterly captivated

Figures
192 *Platanthera bifolia* (L.) L.C.M.
193 *Platanthera chlorantha* (Custer) Rchb.f.

Some European orchids

194

by its strong, very sweet, and yet astringent scent—a scent which is frequently described as being 'reminiscent of goat'.

But if you have once been in contact with the actual 'goat compounds', for example 4-methyloctanoic acid and 4-ethyloctanoic acid, or if you have prepared them yourself, then you will realize that the corresponding note in the scent of *Himantoglossum hircinum* could only be compared, at best, to a first-class 'chèvre'. You might also be reminded of a hot iron and freshly-ironed linen, of certain cyclamen species or of the famous 'wet-dog trillium' *(Trillium erectum)*, a resident of the northeastern states of the USA. Lastly, congratulations to those nature-lovers who manage to identify a vanilla note! As the analysis described on page 226 shows, vanilline is in fact present, if only in rather small quantities.

Since my first contact with this fascinating orchid some 35 years ago, I have rediscovered the very special aspect of its scent in a large number of flowering plants belonging to the most diverse families. I was only too pleased to have been able to reveal its secret in the early summer of 1991. The basic scent is, to

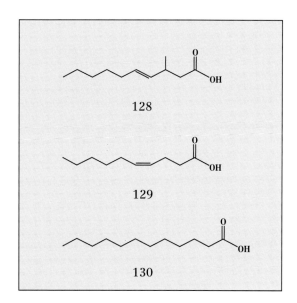

a great extent, attributable to the combination of (E)-3-methyl-4-decenoic acid (128), (Z)-4-decenoic acid (129) and lauric acid (130).

Finally, a series of other compounds, including the ethyl esters of the acids 128 and 129, (E)-ocimene, p-cresol, eugenol and vanilline complete this curious orchid scent (cf. analysis, p. 226).

Figure
194 *Himantoglossum hircinum* (L.) Sprengel

Part Three

The Chemistry of Orchid Scents

General remarks

The third part of this book deals with the chemical and analytical composition of the orchid scents discussed in Part two.

An earlier section (p. 22) has already been devoted to the techniques used for trapping and analysing the orchid scents described in this book. The principle of this analytical approach has already been applied some 20 years ago by e.g. Hills and co-authors and Holman and Heimermann (for references see reviews [12–14] to the investigation of the scents of various orchid species and later on by others to the investigation of many other natural scents; e.g. in 1986 by Mookherjee and co-authors, who compared the composition of the fragrance emitted from living flowers still attached to the plant with those of the picked flowers (for references see reviews [15–16 a]). As a rule, the traps employed contained a thin layer of 5 mg activated charcoal embedded between two grids fused into the wall of a glass tube 65 mm long and 3 mm in diameter. The production and successful use of these filters has been described by Grob [61] and Grob and Zürcher [62].

The results of scent analysis are set out in tables in which the sequence of the individual components corresponds to the order in which they are eluted from a DB-Wax capillary column.

The percentages stated are non-corrected GC integration values relating to the headspace samples obtained by the methods described. They cannot, therefore, be used in this form for reconstitution trials.

Since the composition of the scent of any given species can vary considerably from one plant to another, figures are not generally provided for concentrations below 1%. Apart from those cases in which additional information was thought to be of interest to the reader, such components are indicated in the table with an 'm' (minor compound).

As can be seen from the tables of analytical results reproduced below, many of the scent specimens collected were found to contain (E,E)-2,6-dimethyl-3,5,7-octatrien-2-ol (130) and (E,E)-2,6-dimethyl-1,3,5,7-octatetraene (131) plus the corresponding (E,Z)-isomers at a ratio of approximately 10:1. These compounds were described by Kaiser and Lamparsky [73] as components of the scent of hyacinths some 15 years ago and were also mentioned as being present in a

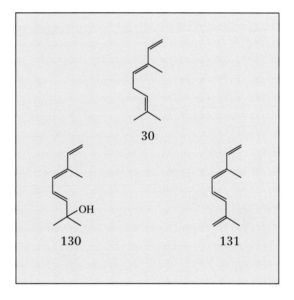

host of other flower scents in a recent literature survey by Kaiser [16a]. According to the most recent findings of Brunke, Hammerschmidt and Schmaus [74 a/b], the presence of the trienol 130 and the tetraene 131 now have to be viewed in a completely new light. Both the alcohol and the fourfold-unsaturated hydrocarbon are apparently produced from the (E)-ocimene already present in the flower scent during the interval between adsorption onto the activated charcoal (the quality of which is of crucial importance to this artefact formation) and elution of the sample collected. The quantity of artefacts formed is determined by the length of sojourn of the ocimene-containing

General remarks

scent adsorbed onto the activated charcoal as well as by its absolute amount, and also perhaps by its qualitative composition. The possibility that 130, 131 and the corresponding (E,Z)-isomers may occasionally be present—at least in small amounts—in the natural scent cannot be ruled out a priori. For the above reasons, the percentages given for the content of (Z)-and (E)-ocimene (30) should be regarded as approximations. They refer to the concentrations, in percentage terms, of the corresponding ocimene, 2,6-dimethyl-3,5,7-octatrien-2-ol and 2,6-dimethyl-1,3,5,7-octatetraene. The following tables do not, therefore, contain any quantitative data for 130 or 131 or their (E,Z)-isomers. The subject of these comments is referred to by means of an asterisk (*).

Whenever a new natural substance is iden-tified in a scent, the table of analytical results is followed by comments relating to the synthesis of the compounds concerned, along with data on their mass spectra. Depending on the complexity of the substance in question, this information is supplemented by NMR and IR data.

In the case of natural substances whose structural elucidation is based on GC/MS investigations and synthesis, identical mass spectra and identical retention behaviour of the synthesized sample on two capillary columns of differing polarities are considered as confirmation of the structure. Whenever the size of the scent sample was adequate, new natural substances were isolated by means of preparative capillary gas chromatography [63] and the NMR spectroscopy subsequently incorporated in the process of structural elucidation. During these investigations it was not possible to establish the absolute configuration of optically active compounds.

GC/MS investigations of scent samples were performed at Givaudan-Roure Research (Dübendorf, Switzerland) under the supervision of Dr J. Schmid.

Characterization of analytical samples

As already mentioned in Part one of this book, the scent of a particular species can vary in terms of both quantitative and qualitative composition not only according to the time of day and maturity of the flower but also, to a considerable degree, according to the plant's genetic make-up. It is therefore worth pointing out once again that these olfactory and analytical investigations were performed on the plants illustrated in Parts one and two of the book.

The tables of analytical results are accompanied by a brief characterization of the analytical sample. This contains, in the following order:
– a code for the analytical investigation in question, e.g. K 13334
– an abbreviation identifying the grower or the location of the plant
  BGSG = Botanical Gardens St. Gallen (Switzerland)
  BGZ = Botanical Gardens Zurich (Switzerland)
  BGHB = Botanical Gardens Heidelberg (Germany)
  Weym. = Dr C. Weymuth, Münsingen, Switzerland
  Kais. = R. Kaiser, Uster, Switzerland
– The date and time of trapping, e.g. Sept. 11, 1988, 11.30–14.30
– The initials of the person performing the trapping experiment
  GG = Günter Gerlach
  CW = Christoph Weymuth
  RK = Roman Kaiser
– a code indicating the size of the sample collected over the period indicated
  a = > 200 µg
  b = 100–200 µg
  c = 10–100 µg
  d = up to 10 µg

An analytical sample may therefore be characterized as follows: K 11334, BGSG, Sept. 11, 1988, 11.30–14.30, RK, a.

Equipment used

Gas chromatography was performed using a Carlo Erba FTV 4160 chromatograph. The following capillary columns were available:

a) UCON HB 5100; 50 m × 0.3 mm, helium as carrier gas (90 kPa). Temperature program 60°–90°C, 2.5°min.
b) Pluronic L64, 50 m × 0.3 mm, helium as carrier gas (85 kPa). Temperature program 50°–200°C, 3°/min.
c) DB-Wx J&W, 30 m × 0.3 mm, helium as carrier gas (70 kPa). Temperature program 50°–210°C, 2.5°/min.

For the GC/MS measurements, a gas chromatograph of the type mentioned above and a Varian MAT Model 212/CH-5 mass spectrometer were used, together with the Finnigan INCOS data system. In the case of mass spectra expressed numerically, the intensities of the molecular ions and the principal fragments are stated as relative intensities in % of the base peak.

A Bruker AM 400 spectrometer was used for NMR investigations. Measurements were performed in $CDCl_3$ with TMS as the internal standard.

Analytical composition of the individual orchid scents

Acineta superba (cf. p. 50)

K 17266, BGHB, April 3, 1991, 9.00–12.00, GG, b.

Compound	Area %
methyl 2-methylbutyrate	m
methyl 2-hydroxy-3-methylvalerate	1.2
isocaryophyllene	1.0
trans-α-bergamotene	6.6
caryophyllene	16.3
(E)-β-farnesene	m
β-bisabolene	3.9
methyl (Z)-cinnamate	m
caryophyllene epoxide	m
methyl (E)-cinnamate	11.1
(Z)-α-*trans*-bergamotenal	2.5
(Z)-α-*trans*-bergamotol	44.5
benzyl acetate	m
methyl phenylacetate	m
(Z,E)-α-farnesene	m
(E,E)-α-farnesene	11.0
cis-linalool oxide (pyranoid)	m
methyl salicylate	m
(E)-geranylacetone	m
benzyl alcohol	1.9
phenylethyl alcohol	m
202(M$^+$,10),159(38),117(24),91(52), 67(88),41(100)	m
(E)-nerolidol	3.8
p-cresol	m
eugenol	m
220(M$^+$,1),162(16),159(12),105(21), 93(29),41(100)	3.2
(E,E)-farnesol	m
indole	m
vanilline	m

* Substitutions at C(1) not indicated.

Aerangis appendiculata (cf. p. 123)

K 17256, BGZH, April 30/May 1, 1991, 20.00–7.00, RK, a.

Compound	Area %
α-pinene	m
β-pinene	m
myrcene	m
heptanal	m
limonene	3.5
isoamyl alcohol	m
p-cymene	m
octanal	m
6-methyl-5-hepten-2-one	m
(Z)-3-hexenol*	m
nonanal	m
trans-linalool oxide (furanoid)	m
cis-linalool oxide (furanoid)	m
decanal	m
isovaleraldoxime (E/Z approx. 2:1)	m
benzaldehyde	m
linalool	49.5
methyl benzoate	15.0
isomenthol	m

Aerangis biloba (cf. p. 124)

K 16719, BGZH, Nov. 15, 1990, 19.00–22.00, RK, a.

Compound	Area %
α-pinene	m
β-pinene	m
limonene	m
isoamyl alcohol	m
nonanal	m
benzaldehyde	8.8
linalool	m
caryophyllene	m
methyl benzoate	25.1
benzyl formate	m
germacrene D	44.0
bicyclogermacrene	3.0
methyl salicylate	1.0
(E)-geranylacetone	m
phenylethyl acetate	m
benzyl alcohol	4.0
phenylethyl alcohol	2.4
germacra-1(10), 5-dien-4-ol	m
p-cresol	1.0
benzyl benzoate	m
benzyl salicylate	m

Aerangis brachycarpa (cf. p. 124)

K 16654, BGZH, Oct. 26, 1990, 20.00–23.00, RK, a/b.

Compound	Area %
α-pinene	m
β-pinene	m
myrcene	m
limonene	8.2
eucalyptol	2.8
6-methyl-5-hepten-2-one	m
(Z)-3-hexenol	m
nonanal	m
decanal	m
p-methyl anisole	4.3
benzaldehyde	8.0
linalool	26.5
linalyl acetate	m
bornyl acetate	m
methyl benzoate	m
phenylacetaldehyde	m
benzyl formate	m
benzyl acetate	3.8
methyl phenylacetate	2.0
methyl salicylate	m
phenylethyl formate	m
phenylethyl acetate	m
(E,E)-4,8,12-trimethyl-1,3,7,11-tridecatetraene	1.8
anethole	6.6
geraniol	m
benzyl alcohol	12.2
phenylethyl alcohol	12.2
(E)-nerolidol	m
methyl (E)-cinnamate	m
p-cresol	m
eugenol	m
methyl isoeugenol	m
(E)-isoeugenol	m
indole	2.3
vanilline	m
p-methoxy cinnamyl acetate	m
benzyl benzoate	m
phenylethyl benzoate	m
p-methoxy cinnamic alcohol	m

Aerangis confusa (cf. p. 125)

K 16653, BGSG, Nov. 8, 1990, 17.30–21.00, RK, a/b.

Compound	Area %
α-pinene	m
butyl acetate	m
myrcene	m
limonene	1.5
eucalyptol	m
(E)-ocimene	m
nonanal	m
methyl 3-methyloctanoate (82)	3.2
menthone	m
decanal	m
benzaldehyde	1.8
linalool	2.2
bornyl acetate	m
methyl benzoate	4.5
menthol	m
benzyl formate	m
germacrene D	m
benzyl acetate	31.2
methyl phenylacetate	m
(Z,E)-4,8,12-trimethyl-1,3,7,11-tridecatetraene	m
(E,E)-4,8,12-trimethyl-1,3,7,11-tridecatetraene	28.0
phenylethyl acetate	1.1
benzyl alcohol	m
phenylethyl alcohol	m
(E,E)-11(12)-epoxy-4,8,12-trimethyl-1,3,7-tridecatriene (132)	m
methyl (E)-2,6,10-trimethyl-5,9-undecadienoate (84)	1.3
(E)-nerolidol	m
methyl (E)-cinnamate	5.2
p-cresol	m
(E)-cinnamyl acetate	m
methyl (E,E)-3,7,11-trimethyl-2,6,10-dodecatrienoate (85)	m
cis-4-methyl-5-decanolide (81)	2.7
methyl (E,E)-4,8,12-trimethyl-3,7,11-tridecatrienoate (86)	6.2
indole	m
benzyl benzoate	1.2
benzyl salicylate	m

cis-4-Methyl-5-decanolide (81, aerangis lactone)

Synthesis:
Aerangis lactone 81 is readily obtainable by hydrogenation of the dihydrojasmone (130) and subsequent Baeyer-Villiger oxidation. Depending on the reaction conditions, 81 is accompanied by its *trans*-isomer 131 in ratio of between 10:1 and 1:1.

Spectral data:
81. -MS: 184 (M$^+$, 0.2), 128(3), 113(23), 99(3), 85(15), 84(39), 69(5), 56(100), 55(24), 43(26),

41(22). ¹H-NMR (400 MHz): 0,90(t, J~7, 3H); 0,96(d, J~7, 3H); 1,25–1,60(m, 6H together); 1,62–1,72(m, 2H); 1,95–2,10(m, 3H); 2,53(d×d, J~7, 2 H); 4,28(m, 1H). -¹³C-NMR: CH₃-signals at 12,60 (C(11)); 14,04 (C(10)). CH₂-signals at 22,58; 25,32; 26,21; 26,85; 31,70; 32, 01. CH-signals at 29, 44(C(4)); 83,07(C(5)). C-signal at 171,96(C(1)). -IR: 1735, 1238, 1200, 1140, 1095, 1069, 1054,993, 908.

131. -MS: 184 (M⁺, 0,2), 128(4), 114(10), 113(100), 99(3), 95(2), 85(27), 84(44), 69(8), 67(8), 57(14), 56(100), 55(34), 43(29), 41(26). -¹H-NMR (400 MHz): 0,90(t, J~7,3H); 1,00(d, J~7,3H); 1,22–1,64(m, 8H); 1,66–1,78(m, 2H); 1,86–1,94(m, 1H); 2,42–2,51(m, Hₐ-C(2)); 2,58–2,65(m, H_b-C(2); 3,95(m, 1H). -¹³C-NMR.: CH₃-signals at 13,96 (C(10)); 17,39(C(11)); CH₂-signals at 22,49; 24,10; 27,74; 29,46; 31,66; 33,40. CH-signals at 32, 17(C(4)); 85,84(C(5)); C-signal at 171,84(C(1)). - IR: 1735, 1248, 1200, 1118, 1098, 1080, 1032, 2020.

Methyl 3-methyloctanoate (82)

Synthesis:
Wittig-Horner reaction of 2-heptanone with methyl P, P-diethylphosphonoacetate and subsequent hydrogenation.

Spectral data:
MS: 172(M⁺, 1), 141(6), 129(1), 115(3), 101(32), 98(4), 87(4), 83(4), 75(24), 74(100), 69(9), 59(13), 57(18), 43(41).

Esters 84–86

Synthesis:
The respective standard reactions are performed on (E)-geranylacetone.
Spectral data:
84. -MS: 236(M⁺, 1), 195(28), 163(19), 137(10), 123(11), 109(92), 95(14), 88(25), 81(27), 69(100), 67(35), 59(8), 55(18), 41(81).

85. -MS: 250(M⁺, 3), 235(1), 219(3), 207(6), 175(3), 147(6), 136(12), 121(27), 114(45), 107(7), 93(9), 81(26), 69(100), 59(5), 41(37).

86. -MS: 264(M⁺, 2), 221(2), 180(2), 153(3), 136(23), 121(36), 107(7), 93(16), 85(12), 81(35), 69(100), 59(6), 55(7), 41(48).

(E,E)-11(12)-Epoxy-4,8,12-trimethyl-1,3,7-tridecatriene (132)

Synthesis:
Epoxidation of (E,E)-11(12)-expoxy-4,8,12-trimethyl-1,3,7,11-tridecatetraene (83, p. 127) and subsequent isolation of the 11(12)-epoxide [16a].

Spectral data:
MS: 234(M⁺, 0,1), 201(1), 173(1), 164(1), 153(5), 148(6), 135(10), 127(6), 107(12), 94(18), 85(13), 81(65), 79(28), 71(70), 59(14), 55(16), 41(100). – ¹H-NMR(400 MHz): 1,27(s, 3H); 1,30(s, 3H); 1,57–1,68(m, 2H); 1,62(s, 3H); 1,76(s, 3H); 2,05–2,18(m, 6H); 2,70(t, J~6,1H); 4,98(bd, J~10,1H); 5,09(bd, J~17,1H); 5,17(m, 1H); 5,85(bd, J~10,1H); 6,58 (t×d, J₁~10, J₂~17,1H).

Aerangis distincta (cf. p. 127)

K 16717, BGZH, Sept. 15/16, 1990, 22.00–7.00, RK, a/b.

Compound	Area %
α-pinene	m
hexanal	m
β-pinene	m
butanol	m
limonene	m
isoamyl alcohol	m
6-methyl-5-hepten-2-one	m
nonanal	m
menthone	m
decanal	m
benzaldehyde	m
linalool	m
methyl benzoate	1.6
methol	m
salicylic aldehyde	m
(E)-β-farnesene	m
benzyl acetate	3.0
(Z,E)-α-farnesene	33.7
(E,E)-α-farnesene	7.9
methyl salicylate	16.0
geranyl acetate	m
phenylethyl acetate	2,6
geraniol	m
benzyl alcohol	m
benzyl isovalerate	m
204(M$^+$, 22), 135(70), 107(100), 93(55), 43(56), 41(55)	m
phenylethyl alcohol	12.0
204(M$^+$, 22), 135(60), 107(100), 93(40), 43(67), 41(55)	m
(E)-nerolidol	m
eugenol	m
M$^+$(?), 109(52), 94(82), 69(98), 43(100), 41(55)	m
(Z)-isoeugenol	m
220(M$^+$, 1), 162(12), 159(10), 93(25), 69(20), 43(100)	1.6
(E)-isoeugenol	7.8
(E,E)-farnesol	m
indole	1.7
vanilline	m
benzyl benzoate	1.0
phenylethyl benzoate	m
benzyl salicylate	1.2
phenylethyl salicylate	m

Aerangis fastuosa (cf. p. 127)

K 17088, BGSG, March 11, 1991, 19.00–22.00, RK, a.

Compound	Area %
α-pinene	m
hexanal	m
β-pinene	m
butanol	m
myrcene	m
limonene	1.0
isoamyl alcohol	m
3-octanone	m
octanal	m
nonanal	m
trans-linalool oxide (furanoid)	m
cis-linalool oxide (furanoid)	m
benzaldehyde	m
decanal	m
linalool	24.5
methyl benzoate	1.0
lavandulol	m
germacrene D	7.2
benzyl acetate	1.0
methyl salicylate	25.0
ethyl salicylate	m
benzyl butyrate	m
benzyl alcohol	25.8
benzyl isovalerate	m
phenylethyl alcohol	m
methyl 2-methoxybenzoate	m
benzyl caproate	m
p-cresol	2.0
chavicol	m
methoxyeugenol	m
benzyl benzoate	2,2
benzyl salicylate	1.2

Aerangis kirkii (cf. p. 128)

K 17204, BGZH, April 15, 1991, 19.30–23.00, RK, a.

Compound	Area %
α-pinene	m
myrcene	m
limonene	1.0
isoamyl alcohol	m
octanal	m
6-methyl-5-hepten-2-one	m
methyl 3-methyloctanoate (82)	1.9
p-methyl anisole	2.8
α-cubebene	m

Aerangis

decanal	m
benzaldehyde	m
β-cubebene	m
linalool	3.3
β-copaene	m
caryophyllene	m
methyl benzoate	m
germacrene D	33.0
benzyl acetate	17.3
δ-cadinene	m
methyl phenylacetate	m
phenylethyl acetate	4.7
(E)-geranylacetone	m
benzyl alcohol	m
phenylethyl alcohol	1.2
germacra-1(10), 5-dien-4-ol	m
(E)-nerolidol	m
p-cresol	1.0
anisyl acetate	m
cis-3-methyl-4-decanolide (134)	m
cis-4-methyl-5-decanolide (81)	26.2
chavicol	m
methyl linoleate	m
vanilline	m

cis-3-Methyl-4-decanolide (134)

Synthesis:
The cis-lactone, together with the trans isomer 133, is obtainable by radical addition of heptanol to methyl crotonate [66].

Spectral data of 134:
MS: 184(M⁺, 0,03), 142(11), 124(7), 115(10), 99(100), 97(18), 71(32), 55(31), 43(37), 41(24).

^{1}H-NMR(200 MHz): 0,90(t~6,3H); 1,02(d, J~7, 3H); 1,22–1,40(m, 8H); 1,40–1,75(m, 2H); 2,20 (m, 1H); 2,50–2,76(2m, 2H); 4,42(m, 1H). cis configuration was established by NEO measurements. -IR: 1780, 1385, 1340, 1295, 1212, 1170, 972, 934.

Aerangis kotschyana (cf. p. 129)

K 16964, Weym., Jan. 19, 1991, 19.00–21.00, CW, b.

Compound	Area %
α-pinene	m
myrcene	m
limonene	m
nonanal	m
decanal	m
benzaldehyde	m
linalool	3.8
caryophyllene	m
germacrene D	33.5
benzyl acetate	16.8
(Z,E)-α-farnesene	m
(E,E)-α-farnesene	1.0
methyl salicylate	m
(E,E)-4,8,12-trimethyl-1,3,7,11-tridecatetraene	10.2
benzyl alcohol	10.1
phenylethyl alcohol	m
methyl (E)-2,6,10-trimethyl-5,9-undecadienoate (84)	m
cinnamic aldehyde	m
germacra-1(10), 5-dien-4-ol	1,6

133   134

2 : 1

Compound	Area %
(E)-nerolidol	m
methyl (E,E)-3,7,11-trimethyl-2,6,10-dodecatrienoate	1.5
indole	m
vanilline	m
benzyl benzoate	m
methyl (E)-cinnamate	m
methyl (E,E)-4,8,12-trimethyl-3,7,11-tridecatrienoate (86)	10.9

Aerangis somalensis (cf. p. 129)

K 17149, BGZH, April 2, 1991, 20.00–22.30, RK, a.

Compound	Area %
α-pinene	m
myrcene	m
limonene	2.1
octanal	m
decanal	m
benzaldehyde	m
linalool	10.6
methyl benzoate	31.5
hotrienol	m
(E)-β-farnesene	1.0
α-terpineol	m
(Z,E)-α-farnesene	m
(E,E)-α-farnesene	m
methyl salicylate	m
geraniol	m
benzyl alcohol	m
phenylethyl alcohol	m
(E)-nerolidol	48.6
benzyl tiglate	m
indole	m
benzyl benzoate	m

Aeranthes grandiflora (cf. p. 130)

K 17429, Weym., May 25/26, 1991, 21.00–6.00, CW, a/b.

Compound	Area %
2,3-pentandione	1.8
3,4-hexandione	m
myrcene	m
limonene	m
octanal	m
2-oxo-3-pentanol	7.5
2-hydroxy-3-pentanone	3.4
M⁺(?), 102(30), 88(69), 87(40), 57(56), 43(100)	m
4-oxo-3-hexanol	10.5
M⁺(?), 102(43), 101(20), 87(34), 57(88), 43(100)	1.9
trans-linalool oxide (furanoid)	m
2,6,6-trimethyl-2-vinyl-tetrahydropyran-5-one	m
cis-linalool oxide (furanoid)	29.8
decanal	m
benzaldehyde	24.6
benzyl acetate	m
hydroquinone dimethyl ether	1.0
trans-linalool oxide (pyranoid)	2.0
cis-linalool oxide (pyranoid)	6.1
1-phenyl-1,2-propandione	m
phenylethyl acetate	m
(E)-geranylacetone	m
benzyl alcohol	m
phenylethyl alcohol	m
phenylacetonitrile	m
phenylglyoxal	m
α-hydroxy acetophenone	m

Aerides crassifolia (cf. p. 144)

K 16147, BGSG, June 14, 1990, 12.30–14.30, RK, b.

Compound	Area %
α-pinene	m
ethyl 2-methylbutyrate	m
limonene	m
6-methyl-5-hepten-2-one	m
nonanal	m
decanal	m
benzaldehyde	m
methyl benzoate	5.9
benzyl acetate	75.0
(E,E)-α-farnesene	m
phenylethyl acetate	1.4
(E)-geranylacetone	m
benzyl 2-methylbutyrate	m
benzyl alcohol	2.4
phenylethyl alcohol	2.3
nonadecane	4.7
1-nonadecene	m
benzyl benzoate	m
benzyl salicylate	m

Aerides fieldingii (cf. p. 144)

K 16095, BGSG, May 25, 1990, 12.00–14.30, RK, a/b.

Compound	Area %
isobutyl acetate	m
2-pentyl acetate	1.2
isoamyl acetate	1.3
isobutyl butyrate	m
myrcene	m
limonene	m
isoamyl alcohol	m
2-pentyl butyrate	m
(Z)-ocimene	m
(E)-ocimene	2.7
p-cymene	m
isoamyl butyrate	m
hexyl acetate	m
3-methylbutyl isovalerate	m
(Z)-4,8-dimethyl-1,3,7-nonatriene	m
(E)-4,8-dimethyl-1,3,7-nonatriene	17.9
6-methyl-5-hepten-2-one	m
nonanal	m
(E,E)-2,6-dimethyl-1,3,5,7-octatetraene	*)
benzaldehyde	m
linalool	1.0
(E)-β-farnesene	1.3
(Z,E)-α-farnesene	m
(E,E)-α-farnesene	12.2
phenylethyl acetate	1.3
(E,Z)-2,6-dimethyl-3,5,7-octatrien-2-ol	*)
(E,E)-2,6-dimethyl-3,5,7-octatrien-2-ol	*)
(E)-nerolidol	40.1
methyl (E,E)-3,7,11,-trimethyl-2,6,10-dodecatrienoate	m
(Z,E)-farnesal	m
methyl palmitate	m
(E,E)-farnesal	1.3
(E)-2(3)-dihydrofarnesol	1.6
(E,E)-farnesol	1.1

Aerides lawrenceae (cf. p. 145)

K 16185, BGSG, June 25, 1990, 12.00–14.30, RK, a.

Compound	Area %
ethyl acetate	m
methyl butyrate	m
methyl 2-methylbutyrate	m
α-pinene	m
2-methyl-3-buten-2-ol	m
butyl acetate	m
myrcene	m
methyl tiglate	m
limonene	m
eucalyptol	m
ethyl caproate	m
6-methyl-5-hepten-2-one	m
hexanol	m
nonanal	m
nerol oxide	m
6-methyl-5-hepten-2-ol	m
methyl 2-hydroxy-3-methylvalerate	m
methyl 2-hydroxy-4-methylvalerate	m
linalool	1.4
methyl benzoate	m
ethyl benzoate	m
neral	1,8
methyl geranate	1.5
geranial	4.0
methyl phenylacetate	2.7
geranyl acetate	m
nerol	41.9
geraniol	35.6
(E)-nerolidol	m
methyl (E)-cinnamate	m
geranyl tiglate	m
(Z,E)-farnesal	m
methyl anthranilate	m
(E,E)-farnesal	4.5
(Z,E)-farnesol	m
(E,E)-farnesol	m
coumarine	m
indole	m

Ancistrochilus rothschildianus (cf. p. 139)

K 16882, BGSG, Jan. 2, 1991, 11.30–15.00, RK, a/b

Compound	Area %
methyl caproate	m
limonene	1.0
ethyl caproate	9.2
octanal	m
2-heptanol	m
6-methyl-5-hepten-2-one	m
nonanal	m
ethyl octanoate	1.5
decanal	m
1-pentadecene	10.1
(Z)-1,8-pentadecadiene	3.3
methyl decanoate	1.0
Z,Z-1,8,11-pentadecatriene (?)	18.0
ethyl decanoate	7.2

Compound	Area %
benzyl acetate	m
ethyl undecanoate	m
methyl laurate	3.8
ethyl laurate	35.9
methyl myristate	m
ethyl tetradecanoate	1.9

Angraecopsis amaniensis (cf. p. 131)

K17152, BGZH, April 2, 1991, 20.00–22.30, RK, a/b.

Compound	Area %
butyl acetate	m
limonene	1.2
hexyl acetate	1.3
(Z)-3-hexenyl acetate	15.2
(Z)-3-hexenol	m
methyl benzoate	8.4
benzyl acetate	54.5
methyl salicylate	m
geranyl acetate	4.0
geraniol	2.1
benzyl alcohol	m
methyl 2-methoxybenzoate	8.1
methyl isoeugenol	m
benzyl benzoate	m

Angraecum aporoides (cf. p. 132)

K 16931, BGSG, Jan. 17, 1991, 17.00–22.00, RK, b.

Compound	Area %
limonene	1.8
octanal	m
nonanal	m
cis-linalool oxide (furanoid)	2.2
benzaldehyde	31.0
linalool	8.0
methyl benzoate	m
benzyl acetate	m
(E,E)-α-farnesene	m
phenylethyl acetate	m
benzyl alcohol	2.0
phenylethyl alcohol	5.0
anis aldehyde	1.5
cinnamic aldehyde	m
(E)-nerolidol	26.0
(E)-cinnamic alcohol	1.4
vanilline	m
benzyl benzoate	3.0
phenylethyl benzoate	9.0

Angraecum bosseri (cf. p. 132)

K 16997, BGZH, Dec. 5, 1990, 19.00–22.00, RK, a.

Compound	Area %
isobutanol	m
isovaleronitrile	m
butanol	m
myrcene	m
limonene	m
isoamyl alcohol	m
(Z)-3-hexenyl acetate	m
hexanol	m
(Z)-3-hexenol	m
acetic acid	m
trans-linalool oxide (furanoid)	m
2-methylbutyraldoxime	0.02
isovaleraldoxime (E/Z ca. 2:1)	0.5
benzaldehyde	m
linalool	75.0
alloaromadendrene	m
2-methylbutyric acid	m
germacrene D	4.4
bicyclogermacrene	m
benzyl acetate	m
geranial	m
δ-cadinene	m
cis-linalool oxide (pyranoid)	m
phenylethyl formate	m
phenylethyl acetate	m
anethole	m
(E)-geranylacetone	m
benzyl alcohol	2.2
phenylethyl alcohol	6.8
anis aldehyde	m
methyl eugenol	m
cinnamic aldehyde	m
germacra-1(10), 5-dien-4-ol	m
p-cresol	m
anisyl acetate	m
(E)-cinnamyl acetate	m
eugenol	1.8
(E)-cinnamic alcohol	m
phenylacetaldoxime	m
chavicol	m
(E)-isoeugenol	m
indole	m
isochavicol	1.5
6-methoxyeugenol	m
vanilline	m
p-methoxy cinnamyl acetate	m

Angraecum eburneum ssp. *eburneum*
(cf. p. 133)

K 16914, BGZH, Jan. 7, 1991, 19.00–22.00, RK, a.

Compound	Area %
isovalderaldehyde	m
hexanol	m
isovaleronitrile	m
isoamyl acetate	2.3
limonene	m
isoamyl alcohol	2.2
isoamyl isovalerate	m
prenol	m
hexanol	m
(Z)-3-hexenol	m
prenyl isovalerate	m
isoamyl tiglate	m
isovaleraldoxime (E/Z approx. 2:1)	9.9
benzaldehyde	12.0
methyl benzoate	6.4
phenylacetaldehyde	3.0
benzyl acetate	34.6
methyl salicylate	m
phenylethyl formate	m
phenylethyl acetate	1.0
butyl benzoate	1.3
benzyl alcohol	8.6
isoamyl benzoate	2.5
phenylethyl alcohol	7.1
α-hydroxy acetophenone	m
isoprenyl benzoate	m
benzyl tiglate	m
phenylglyoxal	m
3-hydroxy-4-phenyl-2-butanone	m
phenylacetaldoxime	1.0
indole	m
benzyl benzoate	m
phenylethyl benzoate	m

Angraecum eburneum ssp. *superbum*
(cf. p. 133)

K 16966, BGZH, Jan. 14, 1991, 18.00–22.00, RK, a.

Compound	Area %
2-methylbutanal	m
2-methylbutyronitrile	1.0
(E)-2-methyl-2-butenal	m
limonene	2.2
2-methylbutanol	m
amyl alcohol	m
(Z)-3-pentenol	2.8
(E)-2-methyl-2-buten-1-ol	3.9
hexanol	m
(Z)-3-hexenol	53.0
heptanol	m
(Z)-4-heptenol	m
2-methylbutyraldoxime	26.8
benzaldehyde	m
isovaleraldoxime (E/Z approx. 2:1)	m
benzyl acetate	m
hydroquinone dimethyl ether	m
methyl salicylate	m
phenylethyl acetate	m
benzyl alcohol	m
phenylethyl alcohol	1.2
cinnamic aldehyde	m
methyl anthranilate	m
(E)-cinnamic alcohol	m
indole	m

Angraecum eichlerianum (cf. p. 134)

K 16937, BGSG, Jan. 17, 1991, 18.00–22.00, RK, a.

Compond	Area %
myrcene	m
limonene	m
6-methyl-5-hepten-2-one	m
trans-linalool oxide (furanoid)	m
cis-linalool oxide (furanoid)	m
benzaldehyde	m
linalool	13.0
methyl benzoate	26.7
trans-linalool oxide acetate (pyranoid)	m
benzyl acetate	36.5
(Z,E)-α-farnesene	7.2
trans-linalooloxide (pyranoid)	1.0
(E,E)-α-farnesene	m
methyl salicylate	4.8
cis-linalool oxide (pyranoid)	m
penylethyl acetate	1.6
benzyl alcohol	m
phenylethyl alcohol	m
phenylacetonitrile	m
(E)-nerolidol	m
methyl anthranilate	m
indole	1.9
benzyl benzoate	1.0
phenylethyl benzoate	m
benzyl salicylate	m

Angraecum sesquipedale (cf. p. 132)

K 13470, BGSG, Febr. 1/2, 1988, 20.00–7.00, RK, a/b.

Compound	Area %
isovaleraldehyde	2.5
isovaleronitrile	3.5
isoamyl acetate	m
limonene	m
isoamyl alcohol	1.0
(Z)-3-hexenol	m
isovaleraldoxime (E/Z approx. 2:1)	34.0
benzaldehyde	1.6
linalool	m
methyl benzoate	17.9
phenylacetaldehyde	m
ethyl benzoate	m
benzyl acetate	1.1
hydroquinone dimethyl ether	m
neral	m
methyl salicylate	m
trans-linalool oxide (pyranoid)	m
cis-linalool oxide (pyranoid)	m
geranial	m
geranyl acetate	m
geraniol	m
benzyl butyrate	m
benzyl isovalerate	m
benzyl alcohol	14.8
phenylethyl alcohol	2.5
β-ionone	m
β-ionone epoxide	m
anis aldehyde	m
(Z)-3-hexenyl benzoate	m
anisyl acetate	m
methyl anthranilate	m
anisyl alcohol	m
(E)-cinnamic alcohol	m
dihydroactinidiolide	m
phenylacetaldoxime	2.0
indole	m
benzyl benzoate	3.0
phenylethyl benzoate	m
p-methoxy cinnamic alcohol	m
benzyl salicylate	1.0

Anguloa clowesii (cf. p. 51)

K 16366, BGSG, Aug. 14, 1990, 14.00–14.30, RK, a.

Compound	Area %
α-pinene	m
hexanal	m
β-pinene	m
limonene	m
eucalyptol	6.5
hydroquinone dimethyl ether	91.5

Bollea coelestis (cf. p. 51)

K 16367, BGSG, Aug. 15, 1990, 11.00–15.00, RK, a.

Compound	Area %
caryophyllene	47.0
β-elemene	m
humulene	1.2
germacrene A	2.5

A) Hg(OAc)$_2$, THF/H$_2$O   B) NaBH$_4$

caryophyllene β-epoxide	1.2
caryophyllene epoxide	5.0
humulene epoxide II	m
caryophyll-5-en-2α-ol (23)	42.3

Caryophyll-5-en-2α-ol (23)

Synthesis:
The tertiary alcohol 23 is formed together with other components on oxymercuration/cyclization/reduction of caryophyllene to caryophyllan-2,6α-oxide (135) [23]. 23 is obtained by means of column chromatography and prep. GC.

The compound 23 can be obtained selectively and in a good yield from the above-mentioned diepoxide 136 [64–65] by reductive cleavage of the 2(12)-epoxy-group to epoxy-alcohol 137 with lithium aluminium hydride and subsequent desepoxidation with the Zn/Cu [23].

Caryophyll-5-en-2β-ol (24), which makes up about 0,1% of the scent of *Gongora cassidea* (cf. p. 77), can be readily obtained by desepoxidation of kobusone (138) to nor-caryophyll-5-en-2-one (139) and subsequent reaction with methylmagnesium idodide [23].

Spectral data:
23. -MS: 222(M⁺, 0.5), 166(5), 161(6), 151(47), 148(23), 123(15), 109(35), 108(36), 98(38),

93(57), 81(69), 71(33), 69(39), 67(35), 59(30), 55(35), 43(100), 41(65).
^{1}H-NMR(400 MHz, CDCl$_3$, 20 °C): 0,95(s, 2×CH$_3$-C(10) und CH$_3$-C(2)); 1,66(bs, CH$_3$-C(6)); 1,20–2,30 (m, broad signals, 13H); 5,18–5,40(bm, H-C(5)). In CDCl$_3$ at 20 °C some signals in the ^{13}C-NMR appear extremely broad and can no longer be interpreted. On the other hand, at an appropriate measuring temperature of 105°C in DMSO, isomerization of the double bond and dehydratization to 21 is too rapid. As a compromise, the readings were taken at 95°C. -^{1}H-NMR(400 MHz, DMSO, 95 °C): 0,94 and 0,95(2×s, 2×CH$_3$-C(10)); 1,08(s, CH$_3$-C(2)); 1,63(s, CH$_3$-C(6)); 1,20–2,30(m, 13H); 5,22(m, H-C(5)). -^{13}CNMR (DMSO, 95°C): CH$_3$-signals at 16,68(CH$_3$-C(6)); 19,38(CH$_3$-C(2)); 22,73 + 22,78 (2×CH$_3$-C(10)); CH$_2$-signals at 22,34(C(4)); 28,91 (C(7)); 37,66; 43,40; CH-signals at 47,74 (C(9)) or C(1)); 51,83 (C(9) or C(1)); 123,09 (C(5)); C-signals at 30,30. -IR(CHCl$_3$): 3590, 3430, 1090, 1055, 950, 875.

24. -MS: 222(M$^+$, 0.3), 204(7), 189(6), 161(7), 151(46), 148(45), 133(24), 123(16), 108(41), 93(41), 81(51), 69(33), 59(13), 55(34), 43(100). For further spectral data, see [23].

Brassavola digbyana (cf. p. 59)

K 12742, BGSG, June 3, 1987, 21.00–24.00, RK, a.

Compound	Area %
α-pinene	m
β-pinene	m
myrcene	m
limonene	m
(Z)-ocimene	m
(E)-ocimene	24.6
rose oxide	m
3-octanol	m
(E,Z)-2,6-dimethyl-1,3,5,7-octatetraene	*)
(E,E)-2,6-dimethyl-1,3,5,7-octatetraene	*)
citronellal	m
benzaldehyde	m
linalool	2.9
methyl benzoate	m
citronellyl acetate	m
(E)-β-farnesene	m
(Z,E)-α-farnesene	m
(E,E)-α-farnesene	1.1
citronellol	41.5
(E,Z)-2,6-dimethyl-3,5,7-octatrien-2-ol	*)
(E,E)-2,6-dimethyl-3,5,7-octatrien-2-ol	*)
(E)-nerolidol	21.0
(E,E)-farnesal	m
(E)-2(3)-dihydrofarnesol	m
(E,E)-farnesol	1.0

Brassavola glauca (cf. p. 59)

K 12491, BGSG, March 2, 1987, 20.00–23.00, RK, a.

Compound	Area %
α-pinene	m
β-pinene	m
myrcene	m
limonene	m
eucalyptol	m
(Z)-ocimene	m
(E)-ocimene	23.0
rose oxide	m
(E,Z)-2,6-dimethyl-1,3,5,7-octatetraene	*)
ethyl octanoate	m
(E,E)-2,6-dimethyl-1,3,5,7-octatetraene	*)
citronellal	m
benzaldehyde	m
linalool	4.5
methyl benzoate	m
ethyl decanoate	m
citronellyl acetate	1.0
(E)-β-farnesene	m
neral	m
(Z,E)-α-farnesene	m
(E,E)-α-farnesene	m
geranial	m
citronellol	35.5
(E,Z)-2,6-dimethyl-3,5,7-octatrien-2-ol	*)
(E,E)-2,6-dimethyl-3,5,7-octatrien-2-ol	*)
geraniol	1.5
benzyl alcohol	m
(E)-nerolidol	24.5
(E,E)-farnesal	m
(E)-2(3)-dihydrofarnesol	m
(E,E)-farnesol	m

Brassavola nodosa (cf. p.58)

K 16491, BGHB, Sept. 19/20, 1990, 23.00–7.30, RK, a.

Compound	Area %
myrcene	m
limonene	m
(Z)-ocimene	1.0
(E)-ocimene	35.4

Compound	Area %
(E)-4,8-dimethyl-1,3,7-nonatriene	m
(E,Z)-2,6-dimethyl-1,3,5,7-octatetraene	*)
(E,E)-2,6-dimethyl-1,3,5,7-octatetraene	*)
benzaldehyde	m
linalool	6.6
β-elemene	m
(E)-β-farnesene	m
M⁺(?), 97(5), 82(97), 81(40), 54(30), 43(100)	m
(E,E)-α-farnesene	m
neryl acetate	m
citronellol	m
M⁺(?), 121(4), 99(20), 85(7), 71(100), 59(19)	m
geranyl acetate	m
(E,Z)-2,6-dimethyl-3,5,7-octatrien-2-ol	*)
(E,E)-2,6-dimethyl-3,5,7-octatrien-2-ol	*)
geraniol	24.1
(E)-nerolidol	24.0
methyl anthranilate	m
(E,E)-farnesol	m
indole	m

Brassavola tuberculata (cf. p. 58)

K 16338, BGHB, Aug. 7/8, 1990, 22.00–8.00, RK, a/b.

Compound	Area %
α-pinene	m
myrcene	m
limonene	m
(E)-ocimene	30.5
6-methyl-5-hepten-2-one	m
hexanol	m
(Z)-3-hexenol	m
(E,Z)-2,6-dimethyl-1,3,5,7-octatetraene	*)
(E,E)-2,6-dimethyl-1,3,5,7-octatetraene	*)
decanal	m
benzaldehyde	3.0
linalool	1.0
methyl benzoate	7.4
benzyl acetate	3.0
methyl salicylate	19.1
phenylethyl formate	m
ethyl salicylate	m
phenylethyl acetate	16.0
(E,E)-4,8,12-trimethyl-1,3,7,11-tridecatetraene	3.5
(E,Z)-2,6-dimethyl-3,5,7-octatrien-2-ol	*)
(E,E)-2,6-dimethyl-3,5,7-octatrien-2-ol	*)
benzyl alcohol	5.6
phenylethyl alcohol	1.5
3-phenylpropyl acetate	m
indole	m
benzyl benzoate	m

Brassia verucosa (cf. p. 60)

K 13748, BGSG, May 3, 1988, 11.00–13.00, RK, b.

Compound	Area %
α-pinene	1.2
sabinene	1.1
myrcene	1.0
limonene	m
eucalyptol	3.5
isoamyl alcohol	6.0
(Z)-3-hexenyl acetate	m
(Z)-3-hexenol	m
trans-sabinene hydrate	m
decanal	m
benzaldehyde	m
trans-linalool oxide (furanoid)	m
cis-linalool oxide (furanoid)	65.0
methyl benzoate	1.2
acetophenone	m
hotrienol	m
ethyl benzoate	m
benzyl acetate	m
verbenone	1.0
α-terpineol	1.5
trans-2-hydroxy-eucalyptol	m
p-methyl acetophenone	1.0
methyl salicylate	4.9
3-phenylpropanal	m
cumin aldehyde	m
geraniol	m
(E)-geranylacetone	m
(E,E)-4,8,12-trimethyl-1,3,7,11-tridecatetraene	1.2
benzyl alcohol	m
phenylethyl alcohol	m
3-phenylpropanol	m
isoamyl benzoate	m
methyl 2-methoxybenzoate	m
methyl (E)-cinnamate	m
isoamyl salicylate	m
benzyl benzoate	m

Bulbophyllum lobbii (cf. p. 146)

K 16597, BGZH, Oct. 16, 1990, 9.30–11.30, RK, b.

Compound	Area %
α-pinene	m
myrcene	m
limonene	1.4
(Z)-ocimene	3.2
(E)-ocimene	26.0
(Z)-4,8-dimethyl-1,3,7-nonatriene	0.1
(E)-4,8-dimethyl-1,3,7-nonatriene	17.9
hexanol	m
(E)-3(4)-epoxy-3,7-dimethyl-1,6-octadiene	m
(E,E)-2,6-dimethyl-1,3,5,7-octatetraene	*)
trans-linalool oxide (furanoid)	3.0
cis-linalool oxide (furanoid)	2.1
(E)-ocimene epoxide	1.0
isovaleraldoxime (E/Z approx. 2:1)	m
linalool	21.2
(E)-2(3)-epoxy-2,6-dimethyl-6,8-nonadiene (102)	m
bornyl acetate	m
caryophyllene	3.9
trans-linalool oxide (pyranoid)	m
(E,E)-α-farnesene	5.2
methyl salicylate	m
trans-linalool oxide (pyranoid)	m
(E,Z)-2,6-dimethyl-3,5,7-octatrien-2-ol	*)
(E,E)-2,6-dimethyl-3,5,7-octatrien-2-ol	*)
phenylacetonitrile	5.8
phenylacetaldoxime	m
indole	1.1
benzyl benzoate	m

(E)-2(3)-Epoxy-2,6-dimethyl-6,8-nonadiene (102)

Synthesis:
Epoxidation of (E)-4,8-dimethyl-1,3,7-nonatriene (100, cf. p. 147) [16a].

Spectral data:
MS: 166(M+, 2), 148(2), 123(12), 108(7), 95 (42), 93(47), 85(45), 81(55), 79(100), 71(75), 67(50), 59(99), 43(51), 41(88), 39(51). -¹H-NMR(400 MHz): 1,27(s, 3H); 1,31(s, 3H); 1,62–1,72(m, 2H); 1,79(s, 3H); 2,12–2,28(m, 2H); 2,71 (t, J~6, 1H); 5,01(bd, J~10, 1H); 5,13(bd, J~17, 1H); 5,89(bd, J~10, 1H); 6,57(t×d, J₁~10, J₂~17, 1H). -IR: 1645, 1600, 1248, 1120, 987, 896.

Catasetum pileatum (cf. p. 83)

K 17662, Weym., Aug. 10, 1991, 9.00–15.00, CW, a.

Compound	Area %
α-pinene	1.2
β-pinene	m
sabinene	m
α-phellandrene	m
limonene	1.0
β-phellandrene	6.5
eucalyptol	3.0
p-cymene	1.2
trans-limonene epoxide	3.4
dihydrocarvone	1.1
carvenone	m
piperitone	m
carvone	1.5
benzyl acetate	m
trans-carvone epoxide	71.0
3-phenylpropyl acetate	m
(E)-cinnamyl acetate	m
germacra-1(10), 5-dien-4-ol	2.5
(E)-cinnamic alcohol	m

Catasetum viridiflavum (cf. p. 83)

K 17772, Weym., Sept. 30, 1991, 8.00–9.30, CW, a.

Compound	Area %
α-pinene	m
β-pinene	m
sabinene	m
limonene	m
eucalyptol	1.6
p-cymene	m
(Z)-3-hexenyl acetate	m
nonanal	m
myrcene epoxide	m
cis-limonene epoxide	m
trans-limonene epoxide	1.7
benzaldehyde	m
carvenone	m
dihydrocarvone	1.0
carvotanacetone	m
piperitone	m
benzyl acetate	28.6
trans-carvone epoxide (47)	56.0
cis-carvone epoxide (48)	0.2
p-cymen-8-ol	m
benzyl alcohol	m
dodecyl acetate	m

3-phenylpropyl acetate	m
cinnamic aldehyde	m
3-phenylpropanol	m
trans-(*trans*-carveol) epoxide (49)	0.5
(E)-cinnamyl acetate	m
2-amino benzaldehyde	m
thymol	m
carvacrol	m
(E)-cinnamic alcohol	m
indole	3.9

cis-Carvone epoxide (48)
(2R*, 5R*)-2(3)-epoxy-2-methyl-5-isopropenyl-cyclohexanone

The *cis*-epoxide 48 (cf. p. 83) is formed as the main component on epoxidation of carvone with hydrogen peroxide in the presence of sodium hydroxide, and is accompanied by about 6% of the *trans*-epoxide 47 (cf. p. 83).

Spectral data:
MS: 166(M+, 0,3), 151(1), 148(1), 137(4), 123(27), 109(15), 95(30), 85(50), 82(20), 81(21), 71(15), 67(54), 55(25), 43(100), 41(45). -¹H-NMR(400 MHz): 1,41(s, 3H); 1,72(bs, 3H); 1,91 (d×d×d, J₁~1,2, J₂~11,1, J₃-14,9, H_ax-C(4)); 2,03(d×d, J₁~11,5, J₂~17,6, H_ax-C(6)); 2,37(m, H_eq-C(4)); 2,59(d×d×d, J₁~1,6, J₂~4,7, J₃~17,8, H_eq-C(6)); 2,72(m, H_ax-C(5)); 3,45(d×d, J₁~1,2, J₂~3,1, H_eq-C(3)); 4,72(bs, 1H); 4,79(bs, 1H). (Numbering based on systematic nomenclature.)

trans-(*trans*-Carveol) epoxide (49)
(1S*, 2S*, 5R*)-2(3)-epoxy-2-methyl-5-isopropenyl-cyclohexanol

Synthesis:
Reduction of *trans*-carvone epoxide with sodium borohydride results in a 1:1 mixture of *trans*-(*cis*-carveol) epoxide (140) and *trans*-(*trans*-carveol) epoxide (49). The pure compounds can be readily obtained by column chromatography in which 49 is eluted first (hexane/ether 5:1).

Spectral data:
49. -MS: 168(M+, 0,4), 150(3), 135(9), 125(12), 110(36), 109(47), 97(39), 95(32), 85(30), 81(31), 71(53), 69(45), 67(48), 55(32), 43(100), 41(58). -¹H-NMR(400 MHz): 1,46(s, 3H); 1,52(m, J₁~13,5, J₂=J₃=J₄~2, 1H); 1,69(bs, 3H); 1,68–1,82(m, 2H); 2,07(m, J₁~15, J₂~6, J₃~5, J₄~2, 1H); 2,38(m, J₁=J₂~12, J₃~6, J₄~2, H_ax-C(5)); 3,11(d, J~5, H-C(3)); 4,18(m, H_eq-C(11)); 4,70(m, 2H). -IR(CHCl₃): 3610, 3460, 1645, 1100, 1085, 1042, 925, 910, 892, 860.

140. -MS: 168(M+, 0,4), 150(2), 135(3), 125(18), 110(23), 109(25), 97(32), 95(32), 85(22), 81(24), 71(55), 69(35), 67(34), 55(32), 43(100), 41(59). -¹H-NMR(400 MHz): 1,33(m, J₁~12, J₂~12, J₃~10, H_ax-C(6)); 1,46(s, 3H); 1,68(bs, 3H); 1,74–1,81(m, 2H); 1,94–2,04(m, 2H); 3,16(d, J~5, H-C(3)); 3,86 (m, J₁~10, J₂~10, J₃~5, H_ax-C(1)); 4,69(m, 2H). -IR(CHCl₃): 3610, 3460, 1645, 1100, 1085, 1042, 925, 910, 892, 860.

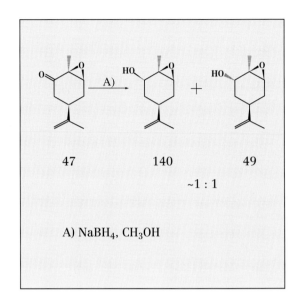

47   140   49

~1 : 1

A) NaBH₄, CH₃OH

Cattleya araguaiensis (cf. p. 66)

K 16490, BGHB, Sept. 19, 1990, 14.30–16.00, RK, b/c.

Compound	Area %
α-pinene	7.0
hexanal	1.5
β-pinene	2.0
sabinene	1.5
myrcene	1.0
heptanal	1.0
limonene	10.5
eucalyptol	2.5
octanal	1.5
6-methyl-5-hepten-2-one	2.0
nonanal	3.5
decanal	5.5
benzaldehyde	m

linalool	2.0
caryophyllene	25.0
methyl benzoate	2.0
hydroquinone dimethyl ether	1.0
methyl salicylate	5.0
(E)-geranylacetone	3.5
caryophyllene β-epoxide	2.0
caryophyllene epoxide	8.0
methyl eugenol	m
eugenol	m
methyl anthranilate	1.0
(Z,E)-farnesal	1.5
(E,E)-farnesal	m
vanilline	m
benzyl benzoate	1.0

Cattleya bicolor (cf. p. 67)

K 16518, BGHB, Sept. 19, 1990, 11.00–14.00, RK, a/b.

Compound	Area %
α-pinene	m
myrcene	m
limonene	1.0
eucalyptol	11.0
(E)-ocimene	6.0
(E,Z)-2,6-dimethyl-1,3,5,7-octatetraene	*)
(E,E)-2,6-dimethyl-1,3,5,7-octatetraene	*)
citronellal	m
benzaldehyde	5.0
6-methyl-5-hepten-2-ol	m
methyl benzoate	15.1
neral	19.0
methyl geranate	m
geranial	9.8
methyl salicylate	2.2
(E,Z)-2,6-dimethyl-3,5,7-octatrien-2-ol	*)
(E,E)-2,6-dimethyl-3,5,7-octatrien-2-ol	*)
(E)-geranylacetone	m
phenylethyl alcohol	m
methyl (Z)-cinnamate	m
methyl (E)-cinnamate	25.0
(E,E)-farnesal	m
(E,E)-farnesol	m
indole	m
benzyl benzoate	m

Cattleya dowiana (cf. p. 63)

K 15397, BGSG, Oct. 27, 1989, 12.30–14.00, RK, a/b.

Compound	Area %
limonene	m
eucalyptol	m
(E)-ocimene	m
6-methyl-5-hepten-2-one	m
ethyl octanoate	m
citronellal	m
benzaldehyde	12.0
linalool	15.5
methyl benzoate	m
neral	12.0
benzyl acetate	m
geranial	15.0
geranyl acetate	m
citronellol	m
phenylethyl formate	m
nerol	2.5
phenylethyl acetate	2.9
geraniol	11.3
benzyl alcohol	7.0
phenylethyl alcohol	17.0
(E)-nerolidol	m
1-nitro-2-phenylethane	m
2-amino benzaldehyde	m
eugenol	m
indole	1.1
vanilline	m
benzyl benzoate	m
phenylethyl benzoate	m

Cattleya labiata (cf. p. 62)

K 13334, BGSG, Nov. 4, 1988, 11.30–14.30, RK, a/b.

Compound	Area %
α-pinene	3.5
α-thujene	m
camphene	m
β-pinene	m
sabinene	m
myrcene	1.1
limonene	8.3
eucalyptol	m
(Z)-ocimene	m
γ-terpinene	m
(E)-ocimene	1.0
p-cymene	m
terpinolene	m

Cattleya

Compound	Area %
6-methyl-5-hepten-2-one	m
(Z)-3-hexenol	m
trans-sabinene hydrate	m
decanal	m
benzaldehyde	1.9
2-methoxy-3-isobutyl pyrazine	m
lilac aldehyde (4 isomers)	m
linalool	16.1
isocaryophyllene	m
caryophyllene	10.2
terpinen-4-ol	m
methyl benzoate	8.0
humulene	m
ethyl benzoate	m
(E)-β-farnesene	1.0
(Z)-3-hexenyl tiglate	m
benzyl acetate	m
verbenone	1.2
α-terpineol	m
β-bisabolene	13.0
(Z,E)-α-farnesene	m
(E,E)-α-farnesene	11.0
methyl salicylate	6.0
lilac alcohol (4 isomers)	m
dihydro-β-ionone	m
3,5-dimethoxy toluene	m
benzylacetone	m
benzyl alcohol	1.1
phenylethyl alcohol	1.0
β-ionone	m
caryophyllene β-epoxide	m
caryophyllene epoxide	2.6
humulene epoxide II	m
(E)-nerolidol	m
benzyl tiglate	m
(Z)-3-hexenyl benzoate	m
(E)-cinnamyl acetate	m
eugenol	1.0
(E)-cinnamic alcohol	m
indole	m
benzyl benzoate	1.2
phenylethyl benzoate	m
benzyl salicylate	m

Cattleya lawrenceana (cf. p. 65)

K 17183, BGSG, April 6, 1991, 12.00–14.30, RK, b/c.

Compound	Area %
α-pinene	2.5
β-pinene	1.5
myrcene	m
heptanal	m
limonene	1.5
(Z)-ocimene	m
(E)-ocimene	44.0
octanal	m
(Z)-3-hexenyl acetate	m
6-methyl-5-hepten-2-one	1.2
nonanal	m
(E)-3(4)-epoxy-3.7-dimethyl-1.6-octadiene	m
(E,Z)-2,6-dimethyl-1,3,5,7-octatetraene	*)
(E,E)-2,6-dimethyl-1,3,5,7-octatetraene	*)
decanal	1.2
benzaldehyde	1.4
linalool	m
caryophyllene	m
methyl benzoate	1.0
neral	m
benzyl acetate	5.9
geranial	m
(E,E)-α-farnesene	3.5
methyl salicylate	3.3
geranyl acetate	m
(E,Z)-2,6-dimethyl-3,5,7-octatrien-2-ol	*)
(E,E)-2,6-dimethyl-3,5,7-octatrien-2-ol	*)
geraniol	18.9
(E)-nerolidol	2.0
methyl 2-methoxybenzoate	3.0
eugenol	m
(E,E)-farnesol	m
indole	m
vanilline	m

Cattleya leopoldii (cf. p. 69)

K 17562, Weym., July 20, 1991, 10.00–16.00, CW, a.

Compound	Area %
α-pinene	m
β-pinene	m
myrcene	m
limonene	m
3-octanone	m
octanal	m
6-methyl-5-hepten-2-one	m
nonanal	m
decanal	m
benzaldehyde	1.0
linalool	m
methyl benzoate	14.0
hydroquinone dimethyl ether	m
methyl salicylate	22.0
(E)-geranylacetone	m
benzyl alcohol	m
methyl (Z)-cinnamate	m
methyl eugenol	3.0

Cattleya

methyl (E)-cinnamate	44.0
methyl 2-methoxybenzoate	m
benzyl tiglate	m
p-methoxy acetophenone	m
2-amino benzaldehyde	m
methyl (E,E)-3,7,11-trimethyl-2,6,10-dodecatrienoate	m
(Z,E)-farnesal	m
methyl anthranilate	1.5
methyl 2-hydroxy-3-phenylpropionate	m
(E,E)-farnesal	m
(Z,E)-farnesol	m
(E,E)-farnesol	m
indole	3.1
vanilline	1.0
benzyl benzoate	1.1
benzyl salicylate	m
gingerone	m

Cattleya luteola (cf. p. 67)

K 16847, BGSG, Dec. 15, 1990, 5.00–7.00, RK, a/b.

Compound	Area %
α-pinene	m
β-pinene	m
myrcene	m
limonene	1.0
(E)-ocimene	m
hexanol	m
(Z)-3-hexenol	m
nonanal	m
decanal	m
α-copaene	4.5
caryophyllene	83.5
humulene	3.0
germacrene D	m
α-selinene	m
phenylethyl acetate	m
(E,E)-4,8,12-trimethyl-1,3,7,11-tridecatetraene	m
phenylethyl alcohol	m
jasmone	m
caryophyllene epoxide	m
germacra-1(10), 5-dien-4-ol	m
(E)-nerolidol	m
benzyl benzoate	m

Cattleya maxima (cf. p. 65)

K 16504, BGHB, Sept. 19, 1990, 10.30–14.00, RK, a/b.

Compound	Area %
α-pinene	m
myrcene	m
limonene	m
eucalyptol	1.4
(Z)-ocimene	1.0
(E)-ocimene	57.0
6-methyl-5-hepten-2-one	m
methyl octanoate	4.6
(E,Z)-2,6-dimethyl-1,3,5,7-octatetraene	*)
(E,E)-2,6-dimethyl-1,3,5,7-octatetraene	*)
benzaldehyde	4.2
linalool	m
methyl decanoate	3.8
methyl benzoate	13.2
benzyl acetate	m
methyl salicylate	m
phenylethyl acetate	m
(E,Z)-2,6-dimethyl-3,5,7-octatrien-2-ol	*)
(E,E)-2,6-dimethyl-3,5,7-octatrien-2-ol	*)
(E)-geranylacetone	m
benzyl butyrate	m
2-amino benzaldehyde	m
methyl anthranilate	m
(E,E)-farnesal	m
(E,E)-farnesol	2.2
indole	m
benzyl benzoate	m

Cattleya percivaliana (cf. p. 65)

K 16927, BGZH, Jan. 15, 1991, 8.30–12.30, RK, b.

Compound	Area %
myrcene	m
limonene	5.6
octanal	m
6-methyl-5-hepten-2-one	m
nonanal	m
isomenthone	m
decanal	m
benzaldehyde	19.0
linalool	m
caryophyllene	2.9
methyl benzoate	2.2
methyl (Z)-4-decenoate	1.7
lilac acid methyl ester	m
germacrene D	12.5
(Z)-4-decenyl acetate	1.5

(E)-4-decenyl acetate	8.5
β-bisabolene	10.0
benzyl acetate	1.4
(E,Z)-2,4-decadienal	m
methyl salicylate	2.9
methyl (E,Z)-2,4-decadienoate	7.3
methyl (E,E)-2,4-decadienoate	1.0
(E,Z)-2,4-decadienyl acetate	1.0
benzyl alcohol	3.9
(E,E)-2,4-decadienyl acetate	m
methyl (E)-cinnamate	4.5
eugenol	m
benzyl benzoate	1.1
benzyl salicylate	1.2

Cattleya porphyroglossa (cf. p. 68)

K 16119, Weym., June 4, 1990, 10.00–14.00, CW, b/c.

Compound	Area %
α-pinene	m
β-pinene	m
myrcene	m
limonene	1.1
(E)-ocimene	4.0
(E,Z)-2,6-dimethyl-1,3,5,7-octatetraene	*)
(E,E)-2,6-dimethyl-1,3,5,7-octatetraene	*)
benzaldehyde	m
linalool	2.8
neral	2.9
benzyl acetate	m
geranial	3.9
hydroquinone dimethyl ether	m
(E,Z)-2,4-decadienal	m
citronellol	1.7
(E,E)-2,4-decadienal	m
nerol	45.3
phenylethyl acetate	m
(E,Z)-2,6-dimethyl-3,5,7-octatrien-2-ol	*)
(E,E)-2,6-dimethyl-3,5,7-octatrien-2-ol	*)
geraniol	23.0
benzyl alcohol	1.2
phenylethyl alcohol	5.9
(E,Z)-2,4-decadienol	m
(E,E)-2,4-decadienol	m
(E)-cinnamic alcohol	m
indole	m
benzyl benzoate	m
benzyl salicylate	m

Cattleya schilleriana (cf. p. 70)

K 17469, BGZH, July 1, 1991, 12.00–14.00, RK, a.

Compound	Area %
myrcene	1.0
limonene	m
(E)-ocimene	m
6-methyl-5-hepten-2-one	m
nonanal	m
trans-linalool oxide (furanoid)	m
decanal	m
benzaldehyde	8.0
linalool	9.9
neral	m
methyl geranate	m
geranial	2.1
geraniol	68.0
benzyl alcohol	1.5
2-amino benzaldehyde	5.9
(Z,E)-farnesol	m
indole	m
vanilline	m
3,4,5-trimethoxy benzaldehyde	m

Caularthron bicornutum (cf. p. 70)

K 17110, BGZH, March 15, 1991, 11.00–14.00, RK, a/b.

Compound	Area %
methyl isobutyrate	m
myrcene	m
limonene	3.0
6-methyl-5-hepten-2-one	3.6
(Z)-3-hexenol	m
decanal	m
benzaldehyde	2.0
linalool	11.0
isobornyl acetate	m
methyl benzoate	m
cyclic β-ionone	m
methyl salicylate	13.3
dihydro-β-ionone	3.2
(Z)-geranylacetone	1.2
α-ionone	m
(E)-geranylacetone	16.9
benzyl alcohol	8.8
β-ionone	8.6
M+(?), 134(47), 119(59), 107(52), 91(42), 43(100)	8.8
cinnamic aldehyde	m
(E,Z)-pseudoionone	1.1

Compound	Area %
methyl (E)-cinnamate	1.0
(E,E)-pseudoionone	5.8
M+(?), 134(23), 119(31), 93(27), 69(17), 43(100)	3.8
benzyl salicylate	m

Chondrorhyncha lendyana (cf. p. 52)

K 16412, BGSG, Aug. 27, 1990, 12.40–15.00, RK, c.

Compound	Area %
α-pinene	4.4
hexanal	5.4
myrcene	m
heptanal	1.6
limonene	2.3
octanal	1.5
(Z)-3-hexenyl acetate	m
6-methyl-5-hepten-2-one	1.0
hexanol	m
nonanal	4.2
(E)-2-octenal	2.1
butyl caproate	m
decanal	6.8
benzaldehyde	2.2
linalool	m
octanol	1.1
caryophyllene	1.0
(Z)-2-octenol	m
(E)-2-octenol	m
hydroquinone dimethyl ether	m
(E,Z)-2,4-decadienal	9.2
(E,E)-2,4-decadienal	0.5
(Z)-4-decenol	3.6
3,5-dimethoxy toluene	38.5
(E)-geranylacetone	2.6
(E,Z)-2,4-decadienol	0.3
2-amino benzaldehyde	m
benzyl benzoate	m

Cirrhaea dependens (cf. p. 72)

K 11873, BGSG, Aug. 1, 1986, 11.00–14.30, RK, b.

Compound	Area %
α-pinene	2.0
myrcene	15.0
limonene	3.5
eucalyptol	9.5
(Z)-ocimene	1.0
(E)-ocimene	55.0
p-cymene	1.0
(E,Z)-2,6-dimethyl-1,3,5,7-octatetraene	*)
(E,E)-2,6-dimethyl-1,3,5,7-octatetraene	*)
linalool	m
caryophyllene	2.5
germacrene D	1.5
hydroquinone dimethyl ether	2.5
(E,Z)-2,6-dimethyl-3,5,7-octatrien-2-ol	*)
(E,E)-2,6-dimethyl-3,5,7-octatrien-2-ol	*)
caryophyllene epoxide	m
eugenol	1.5
vanilline	m

Cirrhopetalum fascinor (cf. p. 149)

K 16929, BGZH, Jan. 16, 1991, 13.00–14.00, RK, c/d.

Compound	Area %
α-pinene	4.5
limonene	9.0
octanal	1.0
nonanal	0.7
acetic acid	5.0
decanal	0.8
caryophyllene	55.0
humulene	3.0
butyric acid	0.7
caryophyllene epoxide	5.0

Cirrhopetalum robustum (cf. p. 148)

K 17178, BGZH, April 8, 1991, 11.45-13.20, RK, c/d.

Compound	Area %
isobutanol	m
2-methylbutyl acetate	m
limonene	1.0
heptanal	m
butanol	1.5
2-methylbutanol	m
acetoin	m
N,N-dimethyl formamide	2.0
6-methyl-5-hepten-2-one	2.5
hexanol	m
nonanal	2.0
N,N-dimethyl acetamide	m
acetic acid	15.0
decanal	2.6
2,5-hexandione	1.8
benzaldehyde	3.2

Cochleanthes

Compound	Area %
propionic acid	3.7
N-methyl acetamide	2.6
γ-butyrolactone	m
butyric acid	3.5
N-methyl formamide	1.5
salicylic aldehyde	2.5
neral	m
2-methylbutyric acid	3.3
linalool	m
geranial	m
valeric acid	1.2
methyl salicylate	4.0
acetamide	m
nerol	7.0
(E)-geranylacetone	1.3
geraniol	3.0
benzyl alcohol	1.0
phenylethyl alcohol	3.5
cinnamic aldehyde	4.9
methyl (E)-cinnamate	m

Cochleanthes aromatica (cf. p. 53)

K 16228, BGSG, June 4, 1990, 12.00–14.30, RK, a/b.

Compound	Area %
α-pinene	1.0
β-pinene	m
myrcene	m
limonene	m
eucalyptol	1.1
6-methyl-5-hepten-2-one	2.2
trans-linalool oxide (furanoid)	m
6-methyl-5-hepten-2-ol	m
decanal	m
benzaldehyde	1.0
linalool	4.0
methyl benzoate	m
acetophenone	m
ethyl benzoate	m
neral	6.8
methyl geranate	4.9
α-terpineol	2.3
neryl acetate	m
benzyl acetate	m
geranial	10.5
butyl benzoate	m
methyl salicylate	m
geranyl acetate	11.8
citronellol	m
nerol	1.0
phenylethyl acetate	1.1
geraniol	36.0
benzyl alcohol	m
phenylethyl alcohol	1.0
2(3)-epoxy-citral	m
6(7)-epoxy-neral	m
(E)-nerolidol	m
eugenol	1.1
(Z,E)-farnesal	m
methyl anthranilate	m
(E,E)-farnesal	m
(E,E)-farnesyl acetate	1.2
(Z,E)-farnesol	m
(E,E)-farnesol	1.5
veratric aldehyde	m
indole	4.8
vanilline	m
benzyl benzoate	m

Cochleanthes discolor (cf. p. 54)

K 13745, BGSG, May 3, 1988, 11.00–13.00, RK, a.

Compound	Area %
isoamyl alcohol	1.0
benzaldehyde	m
linalool	2.5
methyl benzoate	m
β-elemene	1.0
germacrene A (25)	77.0
(E,E)-4,8,12-trimethyl-1,3,7,11-tridecatetraene	m
benzyl alcohol	m
germacra-1(10), 11-dien-5-one (26)	13.0

Germacra-1(10),11-dien-5-one (26) (tentative)

Only about 20 μg of a 70% specimen of this substance—which is very sensitive to isomerization and cyclization—could be isolated. The postulated structure is based on biomimetic considerations and on the ^{1}H-NMR spectrum of this specimen, plus 2D phase-sensitive COSY DQF.

^{1}H-NMR(400 MHz): 0,95(d, J~6, H$_3$C-C(4)); 1,65(m, H$_a$-C(8)); 1,72(s, 2×3H); 2,25(d×d, J$_1$~19, J$_2$~1,5; H$_a$-C(6)); 2,30(m, H-C(4)); 2,55(m, J$_1$~19, J$_2$~6, H$_b$-C(6)); 2,73(m, J$_1$~12, J$_2$~6, H-C(7)); 4,64(bs, 1H); 4,70(bs,1H); 4,94(m, H-C(1)).

MS: 220(M$^+$, 23), 205(8), 202(7), 159(13), 149(12), 138(12), 123(29), 122(81), 107(42), 95(26), 94(28), 81(54), 69(78), 68(76), 67(70), 55(55), 53(42), 41(100).

Cochleanthes marginata (cf. p. 55)

K 16541, Weym., March 8, 1990, 10.00–14.00, CW, b.

Compound	Area %
α-pinene	1.5
β-pinene	m
limonene	1.0
6-methyl-5-hepten-2-one	m
benzaldehyde	3.8
linalool	1.2
benzyl acetate	m
methyl 3-phenylpropionate	1.0
benzyl alcohol	1.0
phenylethyl alcohol	6.8
3-phenylpropyl acetate	m
methyl (Z)-cinnamate	m
3-phenylpropanol	m
methyl (E)-cinnamate	75.7
benzyl benzoate	m
phenylethyl benzoate	m

Coelogyne zurowetzii (cf. p. 150)

K 17115, BGZH, March 17, 1991, 11.00–12.45, RK, b.

Compound	Area %
α-pinene	1.8
β-pinene	1.0
myrcene	1.6
limonene	3.1
eucalyptol	7.2
(Z)-ocimene	m
(E)-ocimene	19.7
(E)-4,8-dimethyl-1,3,7-nonatriene	3.0
6-methyl-5-hepten-2-one	m
(Z)-3-hexenol	m
(E,Z)-2,6-dimethyl-1,3,5,7-octatetraene	*)
(E,E)-2,6-dimethyl-1,3,5,7-octatetraene	*)
trans-linalool oxide (furanoid)	m
cis-linalool oxide (furanoid)	m
benzaldehyde	2.5
linalool	1.0
β-elemene	2.5
methyl benzoate	m
(E)-β-farnesene	m
α-terpineol	11.9
benzyl acetate	m
(Z,E)-α-farnesene	1.5
trans-linalool oxide (pyranoid)	m
(E,E)-α-farnesene	18.5
methyl salicylate	5.3
cis-linalool oxide (pyranoid)	m
(E,E)-4,8,12-trimethyl-1,3,7,11-tridecatetraene	1.1
(E,Z)-2,6-dimethyl-3,5,7-octatrien-2-ol	*)
(E,E)-2,6-dimethyl-3,5,7-octatrien-2-ol	*)
3,5-dimethoxy toluene	1.9
benzyl alcohol	7.3
phenylacetonitrile	m
eugenol	m
methyl anthranilate	m
(E,E)-farnesal	m
(E,E)-farnesol	1.0
indole	m
benzyl benzoate	m
benzyl salicylate	m

Constantia cipoensis (cf. p. 73)

K 16721, Weym., Nov. 12, 1990, 16.30–18.00, CW, c/d.

Compound	Area %
2-methylbutyronitrile	m
methyl 2-methylbutyrate	m
α-pinene	m
β-pinene	m
myrcene	m
limonene	1.8
6-methyl-5-hepten-2-one	m
2-methylbutyraldoxime	m
benzaldehyde	m
linalool	35.0
methyl benzoate	2.5
neral	m
benzyl acetate	10.0
geranial	1.5
hydroquinone dimethyl ether	3.0
geraniol	2.0
dihydro-β-ionone	1.3
(E)-geranylacetone	8.0
benzyl alcohol	1.3
phenylethyl alcohol	m
β-ionone	1.0
6,10-dimethyl-5,9-undecadien-2-ol	m
(E)-nerolidol	m
methyl (E)-cinnamate	3.0
(E)-cinnamic alcohol	m
(E,E)-farnesal	m
benzyl benzoate	5.0
phenylethyl benzoate	m

Coryanthes

Coryanthes leucocorys (cf. p. 74)

K 17912, Weym., Nov. 10, 1991, 10.00–14.00, RK, a.

Compound	Area %
α-pinene	m
myrcene	m
limonene	m
eucalyptol	1.9
benzaldehyde	m
methyl benzoate	m
benzyl acetate	m
methyl salicylate	88.0
dodecyl acetate	2.2
tetradecanal	m
dodecanol	m
tetradecyl acetate	2.0
tetradecanol	m
hexadecyl acetate	m
hexadecanol	m
indole	m

Coryanthes mastersiana (cf. p. 37)

K 16342, BGHB, Aug. 8, 1990, 10.00–11.15, RK, a.

Compound	Area %
isoamyl alcohol	m
benzaldehyde	1.0
benzyl alcohol	m
phenylethyl alcohol	2.0
2-N-methylamino benzaldehyde	92.0
methyl N-methylanthranilate	m
2-amino benzaldehyde	m

Coryanthes picturata (cf. p. 75)

K 16173, BGHB, Sep. 25, 1990, 9.00–11.00, GG, a.

Compound	Area %
α-pinene	13.5
α-thujene	4.5
β-pinene	17.0
sabinene	34.2
myrcene	1.7
limonene	1.3
β-phellandrene	m
eucalyptol	1.1
(E,Z)-1,3,5-undecatriene	m
myrcene epoxide	1.0
trans-limonene epoxide	m
trans-sabinene hydrate	1.4
geranyl acetate	1.1
2-N-methylamino benzaldehyde	10.5
2(3)-epoxy-geranyl acetate (37)	7.7

2(3)-Epoxy-geranyl acetate (37)

Synthesis:
Epoxidation of geranyl acetate.
MS: 212(M$^+$, 0,01), 197(0,2), 152(1), 130(4), 109(13), 95(10), 88(7), 82(34), 69(42), 67(26), 55(20), 43(100), 41(48).

Coryanthes vieirae (cf. p. 76)

K 16570, BGHB, Oct. 5, 1990, 9.00–12.00, GG, a.

Compound	Area %
myrcene	6.9
hexanal	m
eucalyptol	m
(E)-2-pentenal	m
myrcene epoxide	m
150(M$^+$, 0,1), 135(8), 107(15), 93(20), 79(100), 41(70)	m
acetic acid	1.8
linalool	1.0
trans-α-bergamotene	m
ipsdienone (40)	1.5
ipsdienyl acetate (39)	m
4-methyl-2-pentenolide	m
methyl benzoate	m
ipsdienol	77.5
hydroquinone dimethyl ether	m
(E,E)-α-farnesene	m
(E)-2(3)-epoxy-2,6-dimethyl-5,7-octadien-4-one (41)	m
phenylethyl acetate	m
hydroquinone monomethyl ether	m
phenylethyl alcohol	m
(E)-α-farnesene epoxide	3.8
(E)-nerolidol	m
2-heptadecanone	m
indole	m

Ipsdienyl acetate (39)

MS: 194(M$^+$, 0), 134(19), 127(5), 119(29), 105(7), 91(20), 85(100), 79(12), 67(9), 43(63).

(E)-2(3)-Epoxy-2,6-dimethyl-5,7-octadien-4-one (41)

Synthesis:
Formed together with the (Z)-isomer and the diepoxide directly on epoxidation of ipsdienone with hydrogen peroxide in the presence of sodium hydroxide.

Spectral data:
MS: 166(M$^+$, 0,2), 151(1), 123(1), 108(5), 95(100), 79(22), 67(72), 41(46). -^{1}H-NMR(400 MHz): 1,29(s, 3H); 1,44(s, 3H); 2,30(s, 3H); 3,39(s, 1H); 5,54(d, J~10, 1H); 5,74(d, J~17, 1H); 6,36(s, 1H); 6,43(d×d, J$_1$~10, J$_2$~17, 1H).

Cymbidium goeringii (cf. p. 151)

K 17070, Kais., Febr. 28, 1991, 11.00–16.00, RK, b.

Compound	Area %
α-pinene	m
β-pinene	m
sabinene	m
myrcene	m
limonene	6.1
eucalyptol	1.0
octanal	m
(E)-4,8-dimethyl-1,3,7-nonatriene	m
nonanal	m
isomenthone	m
decanal	m
benzaldehyde	m
linalool	m
caryophyllene	m
menthol	m
isomenthol	m
(E)-β-farnesene	1.0
benzyl acetate	m
(Z,E)-α-farnesene	m
(E,E)-α-farnesene	m
phenylethyl acetate	m
(E)-nerolidol	58.0
1,2,4-trimethoxy benzene (108)	m
(E,E)-farnesal	1.1
methyl *trans*-(Z)-jasmonate	0.4
1,2,3,5-tetramethoxy benzene (141)	3.8
(E,E)-farnesol	11.5
methyl *cis*-(Z)-jasmonate (106)	1.6
methyl *trans*-(Z)-dehydrojasmonate (144)	1.1
methyl *cis*-(Z)-dehydrojasmonate (107)	4.3

1,2,3,5-Tetramethoxy benzene (141)

Synthesis:
Baeyer-Villiger-oxidation of 3,4,5-trimethoxy benzaldehyde to 3,4,5-trimethoxy-1-formyloxy benzene [68] with subsequent saponification and methylation.

Spectral data:
MS: 198(77), 183(100), 155(53), 140(30), 125(29), 123(18), 69(29), 59(18), 53(16). -^{1}H-NMR(400 MHz): 3,80(s, 2×3H); 3,86(s, 2×3H); 6,16(s, 2×1H).

Methyl *trans*-(Z)- (144) and methyl *cis*-(Z)-dehydrojasmonate (107)

Synthesis:
After ketalization with ethylene glycol, commercially available methyl jasmonte (*trans*-(Z)- : *cis*-(Z)- ~88:12) was transferred to the aldehyde 142 by means of ozonolysis. Subsequent performance of the Wittig reaction with allyl-triphenyl phosphorane resulted, after deketalization, in a mixture of the dehydrojasmonates 143, 144, 145 and 107 in the ratio 46:44:6:4.

With the help of prep. GC, 143 and 144 were isolated in a pure form, while the methyl *cis*-(Z)-dehydrojasmonate (107) appeared in a mixture with its *cis*-(E)-isomer 145. Equilibration of this 3:2 mixture yielded 143 and 144 in a ratio of 3:2, along with 6% educt.

Spectral data:
107. -MS: 222(M$^+$, 17), 204(22), 193(10), 149(30), 144(32), 131(53), 130(91), 129(35), 107(62), 91(74), 79(100), 67(98), 59(24), 55(52), 41(89). -^{1}H-NMR (400 MHz): 5,15(d, J~10,5, 1H); 5,23(d×d, J$_1$~1,5, J$_2$~17, H); 5,43(d×t, J$_1$~8, J$_2$~10,5, 1H); 6,06(t, J~10,5, 1H); 6,60(d×t×d, J$_1$~1,5, J$_2$~10,5, J$_3$~17, 1H).

143. -MS: 222(M$^+$, 13), 204(10), 193(9), 149(68), 144(36), 131(49), 130(69), 107(66), 105(52), 91(71), 79(89), 67(100), 59(24), 55(52), 41(84). -^{1}H-NMR(400 MHz): 1,51(m, 1H); 1,93(m, 1H); 2,11(m, 1H); 2,20–2,42(m, 6H); 2,65(m, 1H); 3,70(s, 3H); 5,00(d, J~10,1H); 5,12(d, J~17, 1H); 5,60(d×t, J$_1$~7, J$_2$~15, 1H); 6,09(d×d, J$_1$~10, J$_2$~17, 1H); 6,27(d×t, J$_1$~10, J$_2$~17, 1H).

144. -MS: 222(M$^+$, 13), 204(11), 193(11), 149(79), 144(41), 131(55), 130(79), 107(73), 105(59), 91(81), 79(100), 67(98), 59(26), 55(58), 41(85). -^{1}H-NMR

(400 MHz): 1,51(m, 1H); 1,94(m, 1H); 2,12(m, 1H); 2,20–2,42(m, 4H); 2,51(m, 2H); 2,66(m, 1H); 3,70(s, 3H); 5,15(d, J~10,5, 1H); 5,22(d×d, J₁~1,5, J₂~17, 1H); 5,39(d×t, J₁~8, J₂~10,5, 1H); 6,08(t, J~10,5, 1H); 6,63(d×t×d, J₁~1,5, J₂~10,5, J₃~17, 1H).

Dendrobium anosmum (cf. p. 152)

K 17184, BGZH, April 4, 1991, 12.00–13.30, RK, b.

Compound	Area %
ethyl acetate	2.5
α-pinene	m
dimethyldisulfide	m
heptan-2-one	m
limonene	m
6-methyl-5-hepten-2-one	m
hexanol	m
(Z)-3-hexenol	m
nonanal	m
2-nonanone	m
decanal	m
benzaldehyde	1.7
linalool	1.5
2-undecanone	1.8
methyl benzoate	1.5
ethyl benzoate	m
2-undecyl acetate	m
α-terpineol	2.5
benzyl acetate	3.3
methyl salicylate	m
2-tridecanone	14.8
benzylacetone	1.8
(E)-geranylacetone	m
benzyl alcohol	m
2-tridecyl acetate	m
2-pentadecanone	49.5
ethyl tetradecanoate	m
methyl 2-methoxybenzoate	m
2-pentadecyl acetate	m
2-heptadecanone	4.8
2-nonadecanone	m
indole	m
benzyl benzoate	3.1
benzyl salicylate	m

Dendrobium antennatum (cf. p. 153)

K 16544, BGSG, Oct. 1, 1990, 15.00–17.00, RK, b/c.

Compound	Area %
α-pinene	2.5
butyl acetate	m
hexanal	6.0
butanol	3.0
myrcene	m
heptan-2-one	m
heptanal	m
limonene	1.2
butyl butyrate	8.0
octanal	m
6-methyl-5-hepten-2-one	m
hexanol	4.8
(Z)-3-hexenol	m
nonanal	m
butyl caproate	m

decanal	m
benzaldehyde	2.0
linalool	23.0
bornyl acetate	m
neral	m
geranial	3.5
methyl salicylate	m
citronellol	m
geraniol	31.0
butyl benzoate	m
benzyl butyrate	m
benzyl alcohol	3.0
phenylethyl alcohol	1.0
phenylethyl butyrate	m
butyl laurate	1.0

Dendrobium beckleri (cf. p. 154)

K 17134, BGSG, March 21, 1991, 16.00–18.00, RK, a/b.

Compound	Area %
α-pinene	m
β-pinene	m
myrcene	m
limonene	m
eucalyptol	m
(Z)-ocimene	m
(E)-ocimene	26.0
methyl octanoate	m
trans-sabinene hydrate	m
benzaldehyde	5.5
linalool	28.8
methyl decanoate	m
methyl benzoate	m
hotrienol	m
phenylacetaldehyde	2.0
α-terpineol	1.7
lilac alcohol (4 isomers)	1.2
methyl phenylacetate	9.5
methyl salicylate	2.4
168(M$^+$?, 1), 98(18), 83(100), 71(10), 55(43), 43(27)	m
cis-linalool oxide (pyranoid)	m
methyl 3-phenylpropionate	m
benzyl alcohol	5.0
phenylethyl alcohol	m
anis aldehyde	1.0
3,7-dimethyl-1,6-octadien-3,5-diol (146)	3.2
methyl (E)-cinnamate	m
methyl anisate	1.6
methyl p-methoxyphenylacetate	1.8
3-hydroxy-4-phenyl-2-butanone	1.1
vanilline	m
methyl vanillate	m
benzyl benzoate	m

3,7-Dimethyl-1,6-octadien-3,5-diol (146)

MS: 152(M$^+$-18, 0,5), 137(3), 125(3), 109(5), 96(12), 85(57), 83(44), 82(37), 71(42), 68(100), 67(54), 55(38), 43(84), 41(44).

For further spectral data, see [69].

Dendrobium brymerianum (cf. p. 154)

K 16553, BGSG, Oct. 1, 1990, 15.00–17.00, RK, b/c.

Compound	Area %
α-pinene	3.0
myrcene	2.0
limonene	5.5
(E)-ocimene	5.0
6-methyl-5-hepten-2-one	1.8
nonanal	1.6
(E,E)-2,6-dimethyl-1,3,5,7-octatetraene	*)
decanal	3.0
benzaldehyde	4.2
linalool	41.0
caryophyllene	3.0
methyl benzoate	5.0
benzyl acetate	8.3
methyl salicylate	5.0
(E,E)-4,8,12-trimethyl-1,3,7,11-tridecatetraene	2.5
(E,E)-2,6-dimethyl-3,5,7-octatrien-2-ol	*)
(E)-geranylacetone	1.0
indole	m

Dendrobium carniferum (cf. p. 154)

K 16232, BGSG, July 5, 1990, 11.30–14.30, RK, b.

Compound	Area %
ethyl acetate	1.5
ethyl propionate	2.2
α-pinene	m
ethyl butyrate	3.5
methyl crotonate	m
α-phellandrene	m
ethyl crotonate	m
myrcene	25.2
limonene	9.5
β-phellandrene	2.3
(Z)-ocimene	1.9

Dendrobium

Compound	Area %
(E)-ocimene	m
terpinolene	1.4
(E)-4,8-dimethyl-1,3,7-nonatriene	4.6
6-methyl-5-hepten-2-one	m
nonanal	m
trans-linalool oxide (furanoid)	m
citronellal	m
octyl acetate	m
α-copaene	m
decanal	m
linalool	21.0
caryophyllene	14.5
methyl decanoate	m
methyl benzoate	m
ethyl benzoate	m
citronellyl acetate	m
decyl acetate	m
methyl geranate	1.7
α-terpineol	m
dodecanal	m
neral	m
isobutyl benzoate	m
δ-cadinene	m
citronellol	1.2
p-cymen-8-ol	m
geraniol	m
dodecyl acetate	m
isoprenyl benzoate	m
(E)-nerolidol	m
methyl (E)-cinnamate	m
hexyl benzoate	m
hexadecanal	m

Dendrobium chrysotoxum (cf. p. 156)

K 13696, BGSG, April 15, 1988, 12.00–14.00, RK, b.

Compound	Area %
α-pinene	22.9
β-pinene	2.8
sabinene	1.3
myrcene	28.1
limonene	3.7
eucalyptol	m
(Z)-ocimene	m
(E)-ocimene	6.3
hexal acetate	m
(E)-4,8-dimethyl-1,3,7-nonatriene	m
cis-sabinene hydrate	m
octyl acetate	18.9
decanal	m
benzaldehyde	m
linalool	1.5
octanol	1.5
terpinen-4-ol	m
phenylacetaldehyde	m
cis-verbenol	m
decyl acetate	m
verbenone	m
benzyl acetate	m
neryl acetate	m
geranyl acetate	4.9
geranial	m
phenylethyl acetate	m
geraniol	m
ethyl (E)-cinnamate	m
eugenol	m
phenylethyl tiglate	m
dihydroactinidiolide	m
methyl *trans*-(Z)-jasmonate	m
methyl *cis*-(Z)-jasmonate	m

Dendrobium delacourii (cf. p. 158)

K 16186, BGSG, June 25, 1990, 13.00–14.30, RK, b.

Compound	Area %
hexanal	3.0
heptanal	m
(Z)-3-hexenal	7.5
(Z)-2-hexenal	10.1
octanal	m
(Z)-3-hexenyl formate	1.4
6-methyl-5-hepten-2-one	m
hexanol	m
(E)-3-hexenol	1.0
(Z)-3-hexenol	6.8
nonanal	m
3-octanol	m
(Z)-2-hexenol	1.0
(E)-2-hexenol	19.9
1-octen-3-ol	1.8
decanal	1.0
benzaldehyde	4.0
pentadecane	6.0
linalool	2.5
methyl benzoate	3.4
(Z)-5-octenol	1.3
ethyl benzoate	m
benzyl acetate	13.0
(E,E)-α-farnesene	m
methyl nicotate	1.0
geranial	m
phenylethyl acetate	m
(E)-geranylacetone	m
geraniol	1.0

benzyl alcohol	m
phenylethyl alcohol	2.7
nonadecane	2.0
dodecyl acetate	m

Dendrobium lichenastrum (cf. p. 159)

K 17029, BGSG, Febr. 11, 1991, 14.00–16.00, RK, c.

Compound	Area %
α-pinene	3.0
β-pinene	8.0
β-phellandrene	4.0
myrcene	1.3
limonene	17.0
6-methyl-5-hepten-2-one	1.5
nonanal	1.9
perillene	1.0
decanal	3.0
benzaldehyde	2.9
linalool	33.6
methyl benzoate	m
dodecanal	1.3
benzyl acetate	2.9
geranyl acetate	6.6
phenylethyl acetate	m
(E)-geranylacetone	1.3
benzyl alcohol	1.6
phenylethyl alcohol	m
cinnamic aldehyde	1.0
methyl (E)-cinnamate	1.0
vanilline	m
benzyl benzoate	1.1

Dendrobium moniliforme (white) (cf. p. 159)

K 16408, BGSG, Aug. 30, 1990, 12.30–14.30, RK, a/b.

Compound	Area %
α-pinene	m
β-pinene	m
myrcene	m
limonene	m
eucalyptol	m
(Z)-ocimene	m
(E)-ocimene	32.0
6-methyl-5-hepten-2-one	1.7
hexanol	m
(Z)-3-hexenol	m
nonanal	m
(E,Z)-2,6-dimethyl-1,3,5,7-octatetraene	*)
(E,E)-2,6-dimethyl-1,3,5,7-octatetraene	*)
decanal	m
benzaldehyde	m
linalool	20.5
(Z)-3-hexenyl tiglate	m
benzyl formate	m
hydroquinone dimethyl ether	26.0
(E,E)-α-farnesene	m
(E,Z)-2,6-dimethyl-3,5,7-octatrien-2-ol	*)
(E,E)-2,6-dimethyl-3,5,7-octatrien-2-ol	*)
(E)-geranylacetone	m
benzyl alcohol	m
phenylethyl alcohol	8.3
anis aldehyde	m
(E)-nerolidol	m
benzyl tiglate	m
phenylethyl tiglate	m
methyl anthranilate	m
(E,E)-farnesal	m
anisyl alcohol	m
(E,E)-farnesal	3.0
benzyl benzoate	m
phenylethyl benzoate	m

Dendrobium moniliforme (pink) (cf. p. 159)

K 17264, BGSG, April 30, 1991, 12.30–14.45, RK, b/c.

Compound	Area %
α-pinene	2.0
β-pinene	1.2
myrcene	1.8
limonene	1.5
eucalyptol	m
(Z)-ocimene	m
(E)-ocimene	42.0
6-methyl-5-hepten-2-one	m
hexanol	m
(Z)-3-hexenol	m
nonanal	m
(E,Z)-2,6-dimethyl-1,3,5,7-octatetraene	*)
(E,E)-2,6-dimethyl-1,3,5,7-octatetraene	*)
decanal	m
benzaldehyde	m
linalool	4.0
(Z)-3-hexenyl tiglate	m
benzyl formate	m
hydroquinone dimethyl ether	m
(E,E)-α-farnesene	m
(E,Z)-2,6-dimethyl-3,5,7-octatrien-2-ol	*)
(E,E)-2,6-dimethyl-3,5,7-octatrien-2-ol	*)

Dendrobium

Compound	Area %
(E)-geranylacetone	m
benzyl alcohol	22.5
phenylethyl alcohol	10.5
anis aldehyde	m
(E)-nerolidol	m
benzyl tiglate	m
phenylethyl tiglate	m
methyl anthranilate	2.2
(E,E)-farnesal	m
anisyl alcohol	m
(E,E)-farnesol	1.8
benzyl benzoate	1.5
phenylethyl benzoate	m

Dendrobium monophyllum (cf. p. 159)

K 16601, BGZH, Oct. 17, 1990, 15.30–17.30, RK, a.

Compound	Area %
myrcene	1.3
limonene	m
(E)-ocimene	5.0
nonanal	m
(E,Z)-2,6-dimethyl-1,3,5,7-octatetraene	*)
(E,E)-2,6-dimethyl-1,3,5,7-octatetraene	*)
trans-linalool oxide (furanoid)	m
cis-linalool oxide (furanoid)	m
decanal	m
benzaldehyde	m
linalool	52.8
methyl 2,7-dimethyl-6-octenoate	26.0
hotrienol	m
acetaldehyde citronellylethyl acetal	m
M$^+$(?), 150(3), 128(11), 109(10), 84(38), 69(25), 43(100)	1.0
citronellol	1.1
dihydro-β-ionone	m
(E,Z)-2,6-dimethyl-3,5,7-octatrien-2-ol	*)
(E,E)-2,6-dimethyl-3,5,7-octatrien-2-ol	*)
methyl 3-phenylpropionate	1.7
(E)-geranylacetone	1.8
phenylethyl alcohol	m
β-ionone	1.2
methyl (E)-cinnamate	m
methyl 2-hydroxy-3-phenylpropionate	m
indole	m
benzyl benzoate	m

Dendrobium pugioniforme (cf. p. 154)

K 17781, BGSG, April 6, 1991, 14.30–16.15, RK, a/b.

Compound	Area %
α-pinene	1.2
α-thujene	m
β-pinene	1.0
sabinene	1.9
myrcene	1.5
limonene	6.0
eucalyptol	1.6
(E)-ocimene	2.9
terpinolene	m
6-methyl-5-hepten-2-one	m
nonanal	m
trans-sabinene hydrate	1.4
decanal	m
benzaldehyde	15.0
cis-sabinene hydrate	m
linalool	2.5
methyl benzoate	m
α-terpineol	43.0
benzyl acetate	m
(E,E)-α-farnesene	m
(E)-geranylacetone	m
benzyl alcohol	8.3
phenylethyl alcohol	m
phenylacetonitrile	m
anis aldehyde	m
cinnamic aldehyde	m
1-nitro-2-phenylethane	m
3-hydroxy-4-phenyl-2-butanone	m
anisyl alcohol	m
phenylacetaldoxime	2.8
veratric aldehyde	m
vanilline	m
3,4,5-trimethoxy benzaldehyde	1.8
benzyl benzoate	2.0

Dendrobium trigonopus (cf. p. 161)

K 17111, BGSG, March 11, 1991, 12.30–14.00, RK, b/c.

Compound	Area %
α-pinene	2.3
β-pinene	m
sabinene	m
myrcene	2.5
limonene	4.6
eucalyptol	m
(Z)-ocimene	m

Dendrobium

Compound	Area %
(E)-ocimene	33.0
(Z)-3-hexenol	m
(E,Z)-2,6-dimethyl-1,3,5,7-octatetraene	*)
(E,E)-2,6-dimethyl-1,3,5,7-octatetraene	*)
trans-sabinene hydrate	m
benzaldehyde	8.6
cis-sabinene hydrate	m
linalool	1.5
caryophyllene	7.3
methyl benzoate	4.2
phenylacetaldehyde	m
cis-verbenol	1.9
2,6,6-trimethyl-2-cyclohexen-1,4-dione	m
verbenone	7.9
methyl geranate	m
benzyl acetate	m
geranial	3.3
methyl salicylate	m
trans-verbenone epoxide	4.2
(E,Z)-2,6-dimethyl-3,5,7-octatrien-2-ol	*)
(E,E)-2,6-dimethyl-3,5,7-octatrien-2-ol	*)
geraniol	1.2
benzyl alcohol	m
M+(?), 125(12), 112(38), 98(39), 83(57), 43(100)	8.7
methyl (E)-cinnamate	m

Dendrobium unicum (cf. p. 162)

K 17179, BGSG, April 6, 1991, 11.30–13.30, RK, c.

Compound	Area %
α-pinene	m
β-pinene	m
myrcene	m
limonene	m
(E)-ocimene	4.5
6-methyl-5-hepten-2-one	1.6
nonanal	1.5
(Z)-3-hexenol	m
(E,Z)-2,6-dimethyl-1,3,5,7-octatetraene	*)
(E,E)-2,6-dimethyl-1,3,5,7-octatetraene	*)
decanal	m
benzaldehyde	2.0
(E)-3-methyl-4-decenal	1.3
linalool	8.2
octanol	m
methyl salicylate	7.5
(E)-3-methyl-4-decen-1-ol	22.5
(Z)-3-decenol	5.2
(E,Z)-2,6-dimethyl-3,5,7-octatrien-2-ol	*)
(E,E)-2,6-dimethyl-3,5,7-octatrien-2-ol	*)
(E)-geranylacetone	1.0
benzyl alcohol	7.0
phenylethyl alcohol	10.5
cinnamic aldehyde	8.0
3-phenylpropanol	m
methyl (E)-cinnamate	1.5
γ-decalactone	m
2-amino benzaldehyde	m
δ-dodecalactone	1.0
benzyl benzoate	3.0

Dendrobium virgineum (cf. p. 164)

K 16482, BGSG, Sept. 14, 1990, 12.00–14.30, RK, c.

Compound	Area %
myrcene	2.5
limonene	1.9
octanal	1.5
(E)-4,8-dimethyl-1,3,7-nonatriene	3.5
6-methyl-5-hepten-2-one	2.0
nonanal	2.0
decanal	3.5
pentadecane	3.8
linalool	2.8
caryophyllene	45.0
humulene	11.8
benzyl acetate	2.2
(E)-geranylacetone	2.3
phenylethyl alcohol	m
caryophyllene epoxide	3.3
(E)-nerolidol	m

Dendrobium williamsonii (cf. p. 155)

K 15782, BGSG, Febr. 2, 1990, 12.00–14.00, RK, c.

Compound	Area %
α-pinene	1.0
myrcene	1.0
limonene	4.0
octanal	m
nonanal	m
decanal	1.5
benzaldehyde	4.0
linalool	30.0
methyl benzoate	m
ethyl benzoate	m
dodecanal	m
neral	m
(E,E)-α-farnesene	15.0
methyl salicylate	10.0
geranial	m
phenylethyl acetate	m

Dendrochilum, Diaphananthe

Compound	Area %
(E)-geranylacetone	m
benzyl alcohol	m
phenylethyl alcohol	8.0
phenylacetonitrile	m
tetradecanal	8.0
cinnamic aldehyde	m
(E)-nerolidol	1.0
methyl 2-methoxybenzoate	m
ethyl 2-methoxybenzoate	m
(Z)-3-hexenyl benzoate	m
hexadecanal	5.0
eugenol	m
(E,E)-farnesol	1.1
indole	m
benzyl benzoate	m
phenylethyl benzoate	m
benzyl salicylate	1.0

Dendrochilum cobbianum (cf. p. 164)

K 15414, BGSG, Oct. 27, 1990, 12.30–14.00, RK, a.

Compound	Area %
α-pinene	m
β-pinene	m
myrcene	1.1
limonene	m
(Z)-ocimene	1.0
(E)-ocimene	50.0
hexyl acetate	m
(E)-4,8-dimethyl-1,3,7-nonatriene	m
(Z)-3-hexenyl acetate	m
6-methyl-5-hepten-2-one	m
(E)-3(4)-epoxy-3,7-dimethyl-1,6-octadiene	1.2
perillene	m
(E,Z)-2,6-dimethyl-1,3,5,7-octatetraene	*)
(E,E)-2,6-dimethyl-1,3,5,7-octatetraene	*)
(E)-ocimene epoxide	1.1
β-bourbonene	m
benzaldehyde	m
linalool	1.0
caryophyllene	m
methyl benzoate	m
phenylacetaldehyde	m
humulene	m
neral	m
2,2,6-trimethylcyclohex-2-en-1,4-dione	1.0
ipsdienol	m
germacrene D	11.8
benzyl acetate	m
geranial	m
trans-linalool oxide (pyranoid)	m
(Z,E)-α-farnesene	m
(E,E)-α-farnesene	13.5
geranyl acetate	m
phenylethyl formate	m
citronellol	m
phenylethyl acetate	1.8
(E,Z)-2,6-dimethyl-3,5,7-octatrien-2-ol	*)
dihydro-β-ionone	m
(E,E)-2,6-dimethyl-3,5,7-octatrien-2-ol	*)
geraniol	m
benzyl alcohol	m
phenylethyl alcohol	m
β-ionone	m
3,7-dimethyl-1,6-octadien-3,4-diol	m
(E)-nerolidol	m
(E)-2,6-dimethyl-5,7-octadien-2,3-diol	1.8
2-amino benzaldehyde	1.2
vanilline	m
benzyl benzoate	m

Diaphananthe pellucida (cf. p. 134)

K 16849, BGSG, Dec. 14, 1990, 18.00–20.00, RK, b.

Compound	Area %
2-methylbutyronitrile	m
α-pinene	m
isovaleronitrile	3.5
myrcene	m
limonene	2.5
(E)-ocimene	5.7
p-methyl anisole	1,6
(E,Z)-2,6-dimethyl-1,3,5,7-octatetraene	*)
(E,E)-2,6-dimethyl-1,3,5,7-octatetraene	*)
2-methylbutyraldoxime	0.1
isovaleraldoxime (E/Z approx. 2:1)	m
benzaldehyde	2.6
linalool	3.5
benzyl acetate	0.5
(E,E)-4,8,12-trimethyl-1,3,7,11-tridecatetraene	3.3
(E,E)-2,6-dimethyl-3,5,7-octatrien-2-ol	*)
geraniol	1.0
phenylethyl alcohol	1.0
phenylacetonitrile	62.0
phenylacetaldoxime	2.2
benzyl salicylate	2.5

Diaphananthe pulchella (cf. p. 136)

K 16538, BGSG, Oct. 1, 1990, 19.00–21.00, RK, a.

Compound	Area %
α-pinene	m
hexanal	m
limonene	m
hexyl acetate	m
hexanol	m
(Z)-3-hexenol	m
decanal	m
benzaldehyde	m
caryophyllene	10.5
methyl benzoate	33.2
humulene	1.9
benzyl acetate	1.5
methyl salicylate	38.5
phenylethyl acetate	m
benzyl alcohol	1.0
phenylethyl alcohol	1.0
caryophyllene epoxide	m
(E)-nerolidol	5.5
humulene epoxide II	m
(E)-cinnamyl acetate	m
benzyl benzoate	1.0

Dichea rodriguesii (cf. p. 84)

K 17473, Weym., June 24, 1991, 8.00–10.00, CW, b.

Compound	Area %
α-pinene	m
β-pinene	m
hexanal	m
sabinene	m
α-phellandrene	m
estragole	m
limonene	1.5
β-phellandrene	m
eucalyptol	m
p-cymene	1.0
fenchone	m
nonanal	m
(Z)-4-decenal	m
α-terpineol	m
benzyl acetate	49.0
(E,Z)-2,4-decadienal	m
(E,E)-2,4-decadienal	m
phenylethyl acetate	2.6
benzyl alcohol	m
phenylethyl alcohol	m
methyl (E)-cinnamate	2.4

methyl anisate	m
anisyl acetate	32.5
p-methoxy phenylethyl acetate	6.7
anisyl alcohol	m
p-methoxy phenylethyl alcohol	m

Dracula chestertonii (cf. p. 34, 104)

K 16926, BGZH, Jan. 16, 1991, 12.00–14.00, RK, c.

Compound	Area %
3-hexanone	6.5
2-hexanone	3,9
limonene	3.3
octanal	2.3
3-octanone	13.6
1-octen-3-one	7.2
3-octanol	2.6
1-octen-3-ol	46.0
decanal	2.0
octanol	1.2

Dryadella edwallii (cf. p. 104)

K 17118, BGSG, March 11, 1991, 12.00–14.00, RK, c/d.

Compound	Area %
α-pinene	3.0
β-pinene	2.0
sabinene	2.5
myrcene	3.0
limonene	17.0
eucalyptol	2.6
terpinolene	3.5
2-octanone	m
2-nonanone	3.0
nonanal	m
acetic acid	3.5
benzaldehyde	m
trans-α-bergamotene	9.5
sesquiterpene type eremophilene	17.0
1,2-dimethoxy benzene	3.0
phenylethyl alcohol	1.0
anis aldehyde	10.0
creosol	2.5

Embreea, Encyclia

Embreea rodigasiana (cf. p. 81)

K 17307, Weym., May 5, 1991, 13.00–16.00, CW, a.

Compound	Area %
α-pinene	m
β-pinene	m
sabinene	m
myrcene	3.1
limonene	3.5
eucalyptol	42.0
styrene	m
pentadecane	m
myrcene epoxide	m
trans-limonene epoxide	m
trans-sabinene hydrate	m
campher	m
dodecanal	m
hydroquinone dimethyl ether	10.5
tridecanal	m
tetradecanal	32.0
(E)-nerolidol	2.9
(Z,Z,Z)-3,6,9-dodecatrien-1-ol	m

Encyclia adenocarpa (cf. p. 84)

K 16410, BGSG, Aug. 27, 1990, 12.15–15.00, RK, a.

Compound	Area %
α-pinene	m
butyl acetate	m
hexanal	m
β-pinene	m
butanol	m
heptanol	m
limonene	m
butyl butyrate	1.1
butyl 2-methylbutyrate	m
octanal	m
butyl valerate	m
propyl caproate	m
6-methyl-5-hepten-2-one	m
hexanol	m
nonanal	m
butyl caproate	9.7
p-methyl anisole	m
2-methylbutyl caproate	m
butyl (Z)-3-hexenoate	m
decanal	m
benzaldehyde	1.0
amyl caproate	1.0
isoprenyl caproate	m
linalool	2.0
caryophyllene	1.2
prenyl caproate	m
cyclic β-ionone*)	2.2
butyl octanoate	4.5
butyl (x,y)-octenoate	1.1
p-vinyl anisole	1.9
benzyl acetate	m
hydroquinone dimethyl ether	10.2
prenyl octanoate	m
β-elemene	m
methyl salicylate	m
retro-γ-ionone	m
benzyl propionate	m
(E,E)-4,8,12-trimethyl-1,3,7,11-tridecatetraene	m
hexyl octanoate	m
dihydro-β-ionone	m
α-ionone	1.9
butyl benzoate	1.3
benzyl butyrate	8.8
benzyl alcohol	3.5
β-ionone	10.0
benzyl valerate	1.0
benzyl caproate	22.4
benzyl (Z)-3-hexenoate	1.0
2-amino benzaldehyde	1.0
benzyl heptanoate	m
benzyl (E)-2-hexenoate	m
(E,E)-farnesal	m
benzyl caprylate	1.6
(E,E)-farnesol	m
indole	m
benzyl benzoate	1.7

*) 2,5,5,8a-tetramethyl-6,7,8,8a-tetrahydro-5H-1-benzopyrane

Encyclia baculus (cf. p. 86)

K 16116, BGSG, June 1, 1990, 13.00–15.00, RK, a/b.

Compound	Area %
α-pinene	1.1
α-thujene	m
sabinene	m
myrcene	3.9
limonene	1.2
β-phellandrene	m
(Z)-ocimene	m
(E)-ocimene	30.0
6-methyl-5-hepten-2-one	m
p-methyl anisole	m

Encyclia

Compound	Area %
(E,Z)-2,6-dimethyl-1,3,5,7-octatetraene	*)
(E,E)-2,6-dimethyl-1,3,5,7-octatetraene	*)
trans-sabinene hydrate	m
benzaldehyde	10.1
cis-sabinene hydrate	m
linalool	m
methyl benzoate	m
phenylacetaldehyde	m
2,2,6-trimethylcyclohex-2-en-1,4-dione	27.5
verbenone	1.5
α-terpineol	m
neral	m
(E,E)-α-farnesene	1.3
methyl salicylate	2.1
2,6,6-trimethylcyclohexan-1,4-dione	m
phenylethyl acetate	2.3
(E,Z)-2,6-dimethyl-3,5,7-octatrien-2-ol	*)
(E,E)-2,6-dimethyl-3,5,7-octatrien-2-ol	*)
benzyl alcohol	m
benzyl 2-methylbutyrate	m
phenylethyl alcohol	m
creosol	m
5(6)-epoxy-2,2,6-trimethyl-cyclohexan-1,4-dione	m
anis aldehyde	m
(E)-2-methyl-2-buten-1-yl benzoate	m
prenyl benzoate	m
p-cresol	1.9
benzyl tiglate	m
6,10,14-trimethyl-pentadecan-2-one	m
180(M+, 57), 152(61), 137(73), 123(50), 109(80), 91(52)	2.4
3-hydroxy-4-phenyl-2-butanone	2.2
indole	m
vanilline	m
benzyl benzoate	2.6
phenylethyl benzoate	m
benzyl salicylate	m
phenylethyl salicylate	m

Encyclia citrina (cf. p. 86)

K 17490, BGZH, July 8, 1991, 8.30–10.30, RK, a.

Compound	Area %
myrcene	20.5
limonene	m
(E)-ocimene	m
terpinolene	m
6-methyl-5-hepten-2-one	m
methyl octanoate	m
myrcene epoxide	1.0
citronellal	m
decanal	m
benzaldehyde	m
linalool	5.5
methyl 2,7-dimethyl-6-octenoate	m
ipsdienone	14.5
caryophyllene	m
methyl decanoate	1.0
methyl (Z)-4-decenoate	m
neral	2.5
ipsdienol	23.3
methyl geranate	1.1
α-terpineol	m
geranial	17.2
methyl phenylacetate	m
citronellol	3.5
nerol	5.0
geraniol	1.0
benzyl alcohol	m
phenylethyl alcohol	m
prenyl benzoate	m
(Z,E)-farnesal	m
(E,E)-farnesal	m
(E,E)-farnesol	m
indole	m
benzyl benzoate	m
phenylethyl benzoate	m

Encyclia fragrans (cf. p. 87)

K 16414, BGSG, Aug. 27, 1990, 12.30–15.00, RK, a.

Compound	Area %
α-pinene	1.0
α-thujene	m
ethyl butyrate	m
β-pinene	m
sabinene	m
myrcene	m
α-terpinene	m
methyl caproate	m
limonene	m
eucalyptol	m
butyl butyrate	m
ethyl capronate	2.3
(Z)-ocimene	m
γ-terpinene	m
(E)-ocimene	32.0
2-heptyl acetate	m
hexyl acetate	1.0
(E)-4,8-dimethyl-1,3,7-nonatriene	m
(Z)-3-hexenyl acetate	m
2-heptanol	m
6-methyl-5-hepten-2-one	m
hexanol	m

221

Compound	Area %
nonanal	m
2-heptyl butyrate	1.5
butyl caproate	m
hexyl butyrate	5.3
(E,Z)-2,6-dimethyl-1,3,5,7-octatetraene	*)
(E,E)-2,6-dimethyl-1,3,5,7-octatetraene	*)
(Z)-3-hexenyl butyrate	2.0
cis-sabinene hydrate	m
edulan II	m
(E)-ocimene epoxide	m
decanal	m
benzaldehyde	m
terpinen-4-yl acetate	m
linalool	m
octanol	m
(Z)-3,5-hexadienyl butyrate	m
2-heptyl caproate	1.0
edulan I	m
terpinen-4-ol	1.2
methyl benzoate	m
hexyl caproate	2.6
octyl butyrate	m
(Z)-3-hexenyl caproate	1.1
α-terpineol	2.5
benzyl acetate	m
hydroquinone dimethyl ether	8.8
(E,E)-α-farnesene	2.3
methyl salicylate	m
methyl (E,Z)-2,4-decadienoate	m
phenylethyl acetate	m
(E,Z)-2,6-dimethyl-3,5,7-octatrien-2-ol	*)
(E,E)-2,6-dimethyl-3,5,7-octatrien-2-ol	*)
3,5-dimethoxy toluene	10.1
(E)-geranylacetone	m
benzyl butyrate	m
benzyl alcohol	m
β-ionone	m
methyl (Z)-cinnamate	m
methyl (E)-cinnamate	9.2
indole	m
benzyl benzoate	m

Encyclia glumacea (cf. p. 88)

K 16811, BGSG, Dec. 3, 1990, 15.00–17.00, RK, b/c.

Compound	Area %
α-pinene	m
β-pinene	m
myrcene	1.6
limonene	1.0
(E)-ocimene	9.0
hexanol	2.0
(Z)-3-hexenol	m
nonanal	m
(E,Z)-2,6-dimethyl-1,3,5,7-octatetraene	*)
(E,E)-2,6-dimethyl-1,3,5,7-octatetraene	*)
1-octen-3-ol	m
decanal	m
linalool	62.0
caryophyllene	1.9
(E)-β-farnesene	m
(Z)-3-nonen-1-ol	m
α-terpineol	m
lilac aldehyde (4 isomers)	m
lilac alcohol (4 isomers)	2.5
(Z,Z)-3,6-nonadienol	1.0
methyl salicylate	m
phenylethyl acetate	m
(E,Z)-2,6-dimethyl-3,5,7-octatrien-2-ol	*)
(E,E)-2,6-dimethyl-3,5,7-octatrien-2-ol	*)
geraniol	m
phenylethyl alcohol	m
anis aldehyde	3.7
2-amino benzaldehyde	m
anisyl alcohol	m

Epidendrum ciliare (cf. p. 90)

K 16400, Kais., Sept. 1, 1990, 20.10–21.10, RK, a.

Compound	Area %
2-methylbutanal	m
ethyl butyrate	m
butyl acetate	m
2-methylbutyronitrile	1.0
(E)-2-methyl-2-butenal	m
isovaleronitrile	m
butanol	m
myrcene	1.0
limonene	m
2-methylbutanol	m
butyl butyrate	m
ethyl caproate	m
(E)-ocimene	8.0
2-heptyl acetate	m
hexyl acetate	m
(E)-4,8-dimethyl-1,3,7-nonatriene	m
1-nitro-3-methylbutane	m
(Z)-3-hexenyl acetate	m
6-methyl-5-hepten-2-one	m
hexanol	m
(Z)-3-hexenol	m
(E)-3(4)-epoxy-3,7-dimethyl-1,6-octadiene	m
2-heptyl butyrate	m
butyl caproate	m

hexyl butyrate	1.0
(E,Z)-2,6-dimethyl-1,3,5,7-octatetraene	*)
ethyl octanoate	m
(E,E)-2,6-dimethyl-1,3,5,7-octatetraene	*)
trans-linalool oxide (furanoid)	m
(Z)-3-hexenyl butyrate	m
cis-linalool oxide (furanoid)	m
(E)-ocimene epoxide	8.0
2-methylbutyraldoxime	11.5
isovaleraldoxime (E/Z approx. 2:1)	1.0
benzaldehyde	1.6
linalool	45.0
caryophyllene	m
2-heptyl caproate	m
cyclic β-ionone	m
methyl benzoate	2.0
ethyl benzoate	m
neral	m
germacrene D	2.8
bicyclogermacrene	m
benzyl acetate	m
geranial	m
trans-linalool oxide (pyranoid)	m
propyl benzoate	m
(E,E)-α-farnesene	m
methyl salicylate	m
cis-linalool oxide (pyranoid)	m
isobutyl benzoate	m
(E,E)-4,8,12-trimethyl-1,3,7,11-tridecatetraene	5.0
(E,Z)-2,6-dimethyl-3,5,7-octatrien-2-ol	*)
(E,E)-2,6-dimethyl-3,5,7-octatrien-2-ol	*)
butyl benzoate	m
geraniol	m
benzyl alcohol	m
2-methylbutyl benzoate	m
isoamyl benzoate	m
β-ionone	m
cinnamic aldehyde	m
(E)-nerolidol	m
hexyl benzoate	m
benzyl tiglate	m
(Z)-3-hexenyl benzoate	m
methyl anthranilate	m
indole	m
benzyl benzoate	m

Epidendrum lacertinum (cf. p. 91)

K 17414, BGZH, June 17, 1991, 9.15–10.30, RK, b/c.

Compound	Area %
limonene	m
(Z)-ocimene	m
(E)-ocimene	54.0
hexyl acetate	m
(Z)-3-hexenyl acetate	m
hexanol	m
(E,Z)-2,6-dimethyl-1,3,5,7-octatetraene	*)
(E,E)-2,6-dimethyl-1,3,5,7-octatetraene	*)
linalool	1.5
lavandulyl acetate	10.5
germacrene D	26.5
(E,E)-4,8,12-trimethyl-1,3,7,11-tridecatetraene	2.8
(E,Z)-2,6-dimethyl-3,5,7-octatrien-2-ol	*)
(E,E)-2,6-dimethyl-3,5,7-octatrien-2-ol	*)
indole	m

Epidendrum nocturnum (cf. p. 91)

K 16657, BGZH, Oct. 25, 1990, 20.00–23.00, RK, a/b.

Compound	Area %
α-pinene	1.9
sabinene	1.3
myrcene	2.8
limonene	6.2
eucalyptol	73.0
trans-sabinene hydrate	m
cis-sabinene hydrate	m
linalool	m
bornyl acetate	m
terpinen-4-ol	m
alloaromadendrene	m
estragole	m
4-vinyl anisole	m
germacrene D	10.0
bicyclogermacrene	m
anethole	m
methyl eugenol	m
germacra-1(10), 5-dien-4-ol	m
chavicol	m

Epigeneium lyonii (cf. p. 165)

K 17260, BGSG, April 30, 1991, 11.30–14.30, RK, a/b.

Compound	Area %
α-pinene	m
hexanal	m
β-pinene	m
myrcene	m
limonene	4.1

Compound	Area %
octanal	m
6-methyl-5-hepten-2-one	m
nonanal	m
cis-linalool oxide (furanoid)	m
decanal	m
benzaldehyde	m
linalool	1.0
γ-caprolactone	m
γ-octalactone	88.0
γ-decalactone	m

Eria hyacinthoides (cf. p. 166)

K 16347, BGHB, Aug. 8, 1990, 14.00–16.00, RK, c.

Compound	Area %
heptanal	m
limonene	12.0
octanal	1.0
6-methyl-5-hepten-2-one	5.0
chalcogran A	1.5
chalcogran B	1.5
hexanol	2.0
benzyl methyl ether	1.0
nonanal	3.5
decanal	2.0
benzaldehyde	3.0
(Z)-4-decenal	m
linalool	20.0
2,2,6-trimethylcyclohex-2-en-1,4-dione	m
benzyl acetate	4.3
hydroquinone dimethyl ether	10.9
(Z)-4-decenol	1.0
dihydro-β-ionone	3.2
(E)-geranylacetone	2.6
benzyl alcohol	5.8
phenylethyl alcohol	1.6
β-ionone	1.0
dodecanol	5.0
benzyl benzoate	4.0

Gongora armeniaca (cf. p. 77)

K 15299, BGSG, Sept. 25, 1989, 11.45–13.45, RK, a/b.

Compound	Area %
α-pinene	m
myrcene	m
limonene	1.6
eucalyptol	1.1
(Z)-ocimene	m
(E)-ocimene	22.0
6-methyl-5-hepten-2-one	m
(Z)-3-hexenol	m
p-methyl anisole	m
(E,Z)-2,6-dimethyl-1,3,5,7-octatetraene	*)
(E,E)-2,6-dimethyl-1,3,5,7-octatetraene	*)
α-cubebene	m
α-copaene	1.1
(E)-ocimene epoxide	1.0
3-methyl-2-cyclopenten-1-one	m
benzaldehyde	
α-gurjunene	m
β-cubebene	m
linalool	m
caryophyllene	13.1
humulene	3.4
germacrene D	17.5
bicyclogermacrene	2.1
hydroquinone dimethyl ether	6.1
δ-cadinene	1.2
(E,E)-α-farnesene	1.2
(E,Z)-2,6-dimethyl-3,5,7-octatrien-2-ol	*)
(E,E)-2,6-dimethyl-3,5,7-octatrien-2-ol	*)
methyl 3-phenylpropionate	m
phenylacetonitrile	m
(E)-α-farnesene epoxide	1.4
germacra-1(10),5-dien-4-ol	16.9
methyl (E)-cinnamate	m
methyl (E)-p-methoxycinnamate	m

Gongora cassidea (cf. p. 77)

K 16480, BGSG, Sept. 14, 1990, 12.45–14.30, RK, a.

Compound	Area %
2-methylbutanal	m
methyl 2-methylbutyrate	m
2-methylbutyronitrile	m
myrcene	m
(Z)-ocimene	m
(E)-ocimene	31.0
6-methyl-5-hepten-2-one	m
(E,Z)-2,6-dimethyl-1,3,5,7-octatetraene	*)
(E,E)-2,6-dimethyl-1,3,5,7-octatetraene	*)
(E)-ocimene epoxide	m
benzaldehyde	3.0
2-methylbutyraldoxime	m
linalool	22.5
phenylacetaldehyde	m
(E)-β-farnesene	m
phenylacetaldoxime O-methyl ether	m
benzyl acetate	m
hydroquinone dimethyl ether	2.1

(Z,E)-α-farnesene	m
(E,E)-α-farnesene	12.1
(E,Z)-2,6-dimethyl-3,5,7-octatrien-2-ol	*)
(E,E)-2,6-dimethyl-3,5,7-octatrien-2-ol	*)
methyl 3-phenylpropionate	m
(E)-geranylacetone	m
phenylethyl alcohol	m
phenylacetonitrile	1.3
(Z)-α-farnesene epoxide (147)	0.4
caryophyll-5-en-2β-ol (24)	m
(E)-α-farnesene epoxide (44)	13.5
germacra-1(10),5-dien-4-ol	m
(E)-nerolidol	m
methyl (E)-cinnamate	m
elemol	m
1-nitro-2-phenylethane	1.0
p-methoxy acetophenone	m
methyl 2-hydroxy-3-phenylpropionate	1.3
phenylacetaldoxime	3.5
220(M$^+$,1), 162(12), 159(10), 93(25), 69(20), 43(100)	2.5

(E)- and (Z)-α-Farnesene epoxide (44, 147)

Synthesis:
Epoxidation of α-farnesene (E:Z~4:1) [67] and subsequent isolation of the pure isomers 44 and 147 by column chromatography and prep. GC.

Spectral data:
44. -MS: 220(M$^+$, 1), 159(4), 134(56), 119(81), 105(33), 93(74), 91(37), 85(51), 81(62), 80(88), 71(33), 59(56), 55(52), 43(100), 41(85). -^{1}H-NMR(400 MHz): 1,25(s, 3H); 1,30(s, 3H); 1,66(s, 3H); 1,60–1,70(m, 2H); 1,76(s, 3H); 2,08–2,20(m, 2H); 2,70(d, J~6, 1H); 2,85(m, 2H); 4,93(d, J~10, 1H); 5,09(d, J~17, 1H); 5,18(m, 1H); 5,45(m, 1H); 6,36(d×d, J$_1$~10, J$_2$~17, 1H).

147. -MS: 220(M$^+$, 0,1), 159(4), 134(56), 119(71), 105(22), 93(83), 91(24), 85(48), 81(55), 79(48), 71(39), 59(55), 55(48), 43(100), 41(81). -^{1}H-NMR(400 MHz): 1,26(s, 3H); 1,30(s, 3H); 1,66(s, 3H); 1,58–1,70(m, 2H); 1,82(s, 3H); 2,05–2,20(m, 2H); 2,70(d, J~6, 1H); 2,87(m, 2H); 5,10(d, J~10, 1H); 5,18(m, 1H); 5,21(d, J~17, 1H); 5,34(m, 1H); 6,80(d×d, J$_1$~10, J$_2$~17, 1H).

For spectral data on caryophyll-5-en-2β-ol (24) see [23].

Gymnadenia conopea (cf. p. 176)

K 16217, Hostig/CH-Uster, June 30, 1990, 9.00–11.30, RK, a/b.

Compound	Area %
hexanal	m
limonene	m
p-cymene	m
hexyl acetate	m
octanal	m
(Z)-3-hexenyl acetate	m
6-methyl-5-hepten-2-one	1.5
hexanol	m
(Z)-3-hexenol	m
nonanal	m
α,p-dimethylstyrene	m
6-methyl-5-hepten-2-ol	m
decanal	m
benzaldehyde	m
linalool	m
octanol	m
methyl benzoate	m
lavandulyl acetate	m
ethyl benzoate	m
nonanol	m
lavandulol	m
benzyl acetate	45.0
neryl acetate	m
methyl salicylate	m
benzyl isobutyrate	m
geranyl acetate	m
benzyl propionate	m
ethyl salicylate	m
phenylethyl acetate	m
nerol	m
(E)-geranylacetone	m
geraniol	m
benzyl 2-methylbutyrate	m
benzyl butyrate	m
benzyl alcohol	2.6
benzyl valerate	m
phenylethyl alcohol	m
3-phenylpropyl acetate	m
methyl eugenol	5.7
cinnamic aldehyde	m
3-phenylpropanol	m
p-cresol	m
benzyl tiglate	m
(E)-cinnamyl acetate	m
eugenol	3.2
methyl isoeugenol	m
elemicine	3.3
(E)-cinnamic alcohol	1.6
dihydroactinidiolide	m

Compound	Area %
6-methoxyeugenol	m
(E)-isoelemicine	m
vanilline	m
benzyl benzoate	14.5
phenylethyl benzoate	m

Himantoglossum hircinum (cf. p. 179)

K 17392, CH-Bülach, June 9, 1991, 10.00–13.00, RK, b/c.

Compound	Area %
α-pinene	1.0
β-pinene	m
myrcene	m
limonene	5.0
ethyl caproate	m
(Z)-ocimene	m
(E)-ocimene	12.0
hexanol	m
(Z)-3-hexenol	m
nonanal	m
(E,E)-2,6-dimethyl-1,3,5,7-octatetraene	*)
ethyl octanoate	m
decanal	m
linalool	m
ethyl (Z)-4-decenoate	m
(E,Z)-2,6-dimethyl-3,5,7-octatrien-2-ol	*)
(E,E)-2,6-dimethyl-3,5,7-octatrien-2-ol	*)
(E)-geranylacetone	m
p-cresol	m
eugenol	m
elemicine	2,5
(E)-3-methyl-4-decenoic acid	24.0
(Z)-4-decenoic acid	34.0
lauric acid	12.0

Houlletia odoratissima (cf. p. 92)

K 16768, BGHG, Nov. 17, 1990, 9.00–12.00, GG, a.

Compound	Area %
dihydro-β-ionone	m
(E)-geranylacetone	m
β-ionone	10.4
(E,E)-pseudoionone	m
7(11)-epoxy-megastigma-5(7),6(11)-dien-9-one (tentative)	2.4
7(11)-epoxy-megastigma-5(6)-en-9-one	79.0

Huntleya heteroclita (cf. p. 56)

K 16493, BGHB, Sept. 19, 1990, 10.30–14.00, RK, a/b.

Compound	Area %
α-pinene	3.5
β-pinene	m
myrcene	m
limonene	3.5
eucalyptol	2.5
neryl acetate	14.1
geranyl acetate	15.8
nerol	17.5
geraniol	19.7
2(3)-epoxygeranyl acetate	16.9

Huntleya meleagris (cf. p. 56)

K 17281, BGZH, May 9, 1991, 8.15–9.50, RK, a.

Compound	Area %
α-pinene	20.2
camphene	m
β-pinene	2.1
sabinene	1.2
myrcene	2.7
limonene	2.4
β-phellandrene	1.7
eucalyptol	m
3-octanone	m
p-cymene	m
6-methyl-5-hepten-2-one	m
hexanol	m
nonanal	m
α,p-dimethylstyrene	m
heptanol	m
trans-sabinene hydrate	1.1
p-menth-1-en-7-al	m
cis-myrtanal	m
cis-sabinene hydrate	1.0
linalool	m
methyl benzoate	m
myrtenal	m
trans-pinocarveol	m
cis-verbenol	m
trans-verbenol	55.5
verbenone	m
cis-pinocarveol	m
neral	m
trans-p-mentha-1,7-dien-3-ol	m
methyl salicylate	m
citronellol	m

Compound	Area %
myrtenol	2.3
geranial	m
152(M⁺, 3), 119(7), 109(12), 85(100), 67(16), 41(31)	m
p-mentha-1,3-dien-8-ol (148)	m
p-cymen-8-ol	2.1
trans-myrtanol	m
cis-myrtanol	m
perilla alcohol	m
(E,E)-farnesol	m

p-Mentha-1,3-dien-8-ol (148)

Synthesis:
Grignard reaction of 4-methyl-cyclohexa-1,3-dien-1-yl-methyl ketone [70] with methylmagnesium iodide.

Spectral data:
MS: 152(19), 137(36), 134(6), 119(13), 109(37), 95(12), 91(18), 77(10), 67(20), 59(8), 43(100). -¹H-NMR(400 MHz): 1,35(s, 6H); 1,80(s, 3H); 2,10(m, 2H); 2,23(m, 2H); 5,68(m, 1H); 5,87(bd, J~5, 1H).

Laelia albida (cf. p. 92)

K 17024, BGSG, Febr. 11, 1991, 13.00–15.00, RK, a/b.

Compound	Area %
α-pinene	1.0
β-pinene	m
sabinene	m
myrcene	m
limonene	1.5
prenyl acetate	m
6-methyl-5-hepten-2-one	m
hexanol	m
(Z)-3-hexenol	m
decanal	m
benzaldehyde	m
linalool	20.5
caryophyllene	m
methyl benzoate	32.0
acetophenone	m
(Z,Z)-3,6-nonadienyl acetate	m
benzyl acetate	21.7
hydroquinone dimethyl ether	11.2
(Z,Z)-3,6-nonadienol	m
(Z)-geranylacetone	m
(E)-geranylacetone	6.7
(E)-nerolidol	m
methyl (E)-cinnamate	m
1,2,4-trimethoxy benzene	m
(E,E)-pseudoionone	m
(E,E)-farnesyl acetate	m
benzyl benzoate	m

Laelia anceps (cf. p. 93)

K 17025, BGSG, Febr. 12, 1991, 13.30–15.30, RK, b.

Compound	Area %
α-pinene	1.5
hexanal	m
β-pinene	m
sabinene	m
myrcene	2.2
limonene	1.0
eucalyptol	m
(Z)-ocimene	1.0
(E)-ocimene	87.0
(E)-3(4)-epoxy-3,7-dimethyl-1,6-octadiene	m
(E,Z)-2,6-dimethyl-1,3,5,7-octatetraene	*)
(E,E)-2,6-dimethyl-1,3,5,7-octatetraene	*)
(E)-ocimene epoxide	m
decanal	m
benzaldehyde	m
methyl benzoate	m
(E,Z)-2,6-dimethyl-3,5,7-octatrien-2-ol	*)
(E,E)-2,6-dimethyl-3,5,7-octatrien-2-ol	*)
(E)-geranylacetone	m
(E)-nerolidol	m

Laelia autumnalis (cf. p. 94)

K 16881, BGSG, Jan. 2, 1991, 11.30–15.00, RK, a.

Compound	Area %
α-pinene	1.0
β-pinene	m
myrcene	m
heptan-2-one	m
limonene	m
(E)-2-methyl-2-buten-1-yl acetate	m
hexyl acetate	m
(Z)-3-hexenyl acetate	m
6-methyl-5-hepten-2-one	m
hexanol	m
(Z)-3-hexenol	m
decanal	m
benzaldehyde	m
caryophyllene	6.0
methyl benzoate	m

Laelia, Liparis

Compound	Area %
acetophenone	3.0
humulene	m
decyl acetate	m
benzyl acetate	62.5
hydroquinone dimethyl ether	12.9
methyl salicylate	2.0
(Z,Z)-3,6-decadienyl acetate	1,0
(E,Z)-2,4-decadienal	0.6
(E,E)-2,4-decadienal	0.1
phenylethyl acetate	m
methyl laurate	m
(E)-geranylacetone	m
benzyl alcohol	m
dodecyl acetate	m
(E,Z)-2,4-decadienyl acetate	1.0
3-phenylpropyl acetate	m
caryophyllene epoxide	m
cinnamic aldehyde	1.7
methyl (E)-cinnamate	m
tetradecyl acetate	1.1
(E)-cinnamyl acetate	m
anisyl acetate	1.5
methyl palmitate	m
hexadecyl acetate	m
benzyl benzoate	m
benzyl salicylate	m

Laelia gouldiana (cf. p. 95)

K 13364, BGSG, Jan. 2, 1988, 11.00–14.30, RK, a.

Compound	Area %
α-pinene	m
myrcene	m
isoamyl acetate	m
limonene	m
prenyl acetate	m
eucalyptol	m
hexyl acetate	m
(Z)-3-hexenyl acetate	m
hexanol	m
(Z)-3-hexenol	m
decanal	m
acetophenone	2.5
benzyl formate	m
benzyl acetate	51.6
hydroquinone dimethyl ether	25.8
caryophyllene	7.3
methyl salicylate	m
citronellyl acetate	m
humulene	m
decyl acetate	m
geranyl acetate	m
phenylethyl acetate	1.0
dihydro-β-ionone	m
(E)-geranylacetone	m
methyl laurate	m
3-phenylpropyl acetate	m
cinnamic aldehyde	m
β-ionone	m
dodecanal	m
caryophyllene epoxide	1.0
anisyl acetate	m
methyl myristate	m
tetradecyl acetate	1.2

Laelia perinii (cf. p. 95)

K 16596, BGZH, Oct. 17, 1990, 11.30–13.30, RK, c/d.

Compound	Area %
limonene	7.2
3-octanone	5.2
octanal	m
6-methyl-5-hepten-2-one	3.5
nonanal	m
3-octanol	m
1-octen-3-ol	2.2
decanal	1.5
benzaldehyde	m
linalool	18.0
octanol	1.5
methyl benzoate	5.0
benzyl acetate	6.5
methyl salicylate	2.5
phenylethyl acetate	m
(E)-geranylacetone	5.5
benzyl alcohol	1.5
phenylethyl alcohol	2.5
(E)-2-methyl-2-buten-1-yl benzoate	8.5
elemicine	7.2
(E)-isoelemicine	15.0

Liparis viridiflora (cf. p. 166)

K 16746, BGZH, Nov. 21, 1990, 13.45–15.45, RK, b.

Compound	Area %
α-pinene	1.1
hexanal	m
myrcene	m
limonene	1.3
(E)-7-methyl-1,6-dioxaspiro(4,5)decane	8.1
chalcogran A	0.5
chalcogran B	0.5

Compound	Area %
hexanol	m
(Z)-7-methyl-1,6-dioxaspiro(4,5)decane	0.3
benzaldehyde	1.8
(E)-2-nonenal	m
linalool	1.5
(E,Z)-2,6-nonadienal	1.0
methyl benzoate	1.0
(E)-β-farnesene	m
benzyl acetate	m
(Z,E)-α-farnesene	m
(E,E)-α-farnesene	65.0
(E,E)-4,8,12-trimethyl-1,3,7,11-tridecatetraene	m
benzyl alcohol	1.5
methyl (E)-cinnamate	m
benzyl tiglate	m
220(M+, 1), 162(12), 159(10), 93(25), 69(20), 43(100)	3.0
elemicine	m
benzyl benzoate	m

Lycaste aromatica (cf. p. 96)

K 16229, BGSG, July 4, 1990, 11.30–14.30, RK, a/b.

Compound	Area %
α-pinene	m
α-thujene	m
β-pinene	m
sabinene	m
myrcene	m
α-terpinene	m
limonene	m
eucalyptol	m
(Z)-ocimene	m
(E)-ocimene	36.0
p-cymene	m
6-methyl-5-hepten-2-one	m
(E,Z)-2,6-dimethyl-1,3,5,7-octatetraene	*)
(E,E)-2,6-dimethyl-1,3,5,7-octatetraene	*)
trans-sabinene hydrate	m
(Z)-ocimene epoxide	m
decanal	m
benzaldehyde	m
cis-sabinene hydrate	m
linalool	m
6-methyl-3,5-heptadien-2-one	m
terpinen-4-ol	m
methyl benzoate	m
benzyl acetate	m
hydroquinone dimethyl ether	3.3
3-phenylpropanal	m
phenylethyl acetate	m
(E,Z)-2,6-dimethyl-3,5,7-octatrien-2-ol	*)
(E,E)-2,6-dimethyl-3,5,7-octatrien-2-ol	*)
methyl 3-phenylpropionate	1.3
guaiacol	m
geraniol	m
(Z)-cinnamic aldehyde	0.3
phenylethyl alcohol	m
methyl (Z)-cinnamate	0.4
(E)-cinnamic aldehyde	3.8
methyl (E)-cinnamate	45.5
eugenol	1.4
(E)-cinnamic alcohol	m
indole	m
methyl (E)-p-methoxycinnamate	m
benzyl benzoate	m

Lycaste cruenta (cf. p. 96)

K 16319, Weym., Aug. 1, 1990, 11.00–17.00, CW, a/b.

Compound	Area %
α-pinene	2.8
ethyl butyrate	m
hexanal	m
sabinene	m
myrcene	1.0
eucalyptol	m
(E)-ocimene	44.0
6-methyl-5-hepten-2-one	m
(E,Z)-2,6-dimethyl-1,3,5,7-octatetraene	*)
(E,E)-2,6-dimethyl-1,3,5,7-octatetraene	*)
(E)-ocimene epoxide	1.9
decanal	m
benzaldehyde	m
linalool	12.2
methyl benzoate	m
1,2-dimethoxy benzene	m
hydroquinone dimethyl ether	1.9
methyl salicylate	m
(E,Z)-2,6-dimethyl-3,5,7-octatrien-2-ol	*)
(E,E)-2,6-dimethyl-3,5,7-octatrien-2-ol	*)
anis aldehyde	m
methyl (E)-cinnamate	m
methyl p-methoxy-phenylpropionate	0.6
methyl (Z)-p-methoxycinnamate	18.5
methyl (E)-p-methoxycinnamate	6.9

Lycaste, Masdevallia

Lycaste locusta (cf. p. 97)

K 16115, BGSG, June 1, 1990, 13.00–15.00, RK, a/b.

Compound	Area %
ethyl acetate	75.0
ethyl propionate	m
propyl acetate	m
isobutyl acetate	m
ethyl butyrate	3.1
ethyl 2-methylbutyrate	m
butyl acetate	1.3
diethyl carbonate	15.0
isoamyl acetate	m
butyl butyrate	m
ethyl caproate	1.0
hexyl acetate	m
6-methyl-5-hepten-2-one	m
nonanal	m
decanal	m
ethyl benzoate	m
α-ionone	m
β-ionone	m

Masdevallia caesia (cf. p. 99)

K 16934, BGSG, Jan. 17, 1991, 15.00–17.00, RK, c.

Compound	Area %
ethyl 2-methylbutyrate	m
α-pinene	1.0
isobutanol	m
myrcene	m
butanol	1.0
limonene	2.3
isoamyl alcohol	m
(E)-2-methyl-2-buten-1-yl acetate	m
amyl alcohol	m
octanal	m
(E)-2-methyl-2-buten-1-yl isobutyrate	2.5
(E)-2-methyl-2-buten-1-ol	m
6-methyl-5-hepten-2-one	m
hexanol	m
nonanal	m
(E)-2-methyl-2-buten-1-yl 2-methylbutyrate	3.0
acetic acid	2.0
(E)-2-methyl-2-buten-1-yl tiglate	m
decanal	m
benzaldehyde	7.0
α-bergamotene	3.1
caryophyllene	4.9
methyl benzoate	1.2
butyric acid	3.5
ethyl benzoate	1.5
isovaleric acid	0.9
benzyl acetate	6.6
ar-curcumene	1.0
phenylethyl formate	m
phenylethyl acetate	m
benzyl alcohol	2.8
phenylethyl alcohol	8.9
phenylglyoxal	20.0
α-hydroxy acetophenone	8.0
vanilline	m
benzyl benzoate	m

Masdevallia estradae (cf. p. 101)

K 16855, BGSG, Dec. 15, 1990, 7.00–9.00, RK, c.

Compound	Area %
α-pinene	1.5
β-pinene	m
limonene	6.0
eucalyptol	3.0
nonanal	m
decanal	m
linalool	16.0
caryophyllene	5.5
(E)-β-farnesene	6.0
citronellol	m
geraniol	m
benzyl alcohol	1.0
phenylethyl alcohol	1.5
(E)-nerolidol	3.5
(Z,E)-farnesol	m
methyl *trans*-(Z)-jasmonate	0.8
(E,E)-farnesol	30.0
methyl *cis*-(Z)-jasmonate	11.0

Masdevallia glandulosa (cf. p. 101)

K 16853, BGSG, Dec. 15, 1990, 10.00–11.00, RK, c.

Compound	Area %
α-pinene	1.8
β-pinene	1.0
myrcene	2.5
limonene	3.5
(E)-ocimene	11.0
(E,Z)-2,6-dimethyl-1,3,5,7-octatetraene	*)
(E,E)-2,6-dimethyl-1,3,5,7-octatetraene	*)
methyl 3-hydroxybutyrate	10.8
nonanal	m
benzaldehyde	3.1

Compound	Area %
linalool	1.5
benzyl acetate	m
(E,Z)-2,6-dimethyl-3,5,7-octatrien-2-ol	*)
(E,E)-2,6-dimethyl-3,5,7-octatrien-2-ol	*)
benzyl alcohol	14.0
phenylethyl alcohol	6.3
phenylacetonitrile	m
cinnamic aldehyde	3.5
caprylic acid	2.5
(E)-cinnamyl acetate	2.2
eugenol	13.5
(E)-cinnamic alcohol	3.9
chavicol	1.5
(E,E)-farnesol	3.0
vanilline	m
benzyl benzoate	2.8

Masdevallia laucheana (cf. p. 103)

K 16792, BGSG, Dec. 3, 1990, 16.00–17.30, RK, b.

Compound	Area %
myrcene	m
limonene	m
(Z)-ocimene	1.0
(E)-ocimene	60.0
(E,Z)-2,6-dimethyl-1,3,5,7-octatetraene	*)
(E,E)-2,6-dimethyl-1,3,5,7-octatetraene	*)
citronellal	m
neral	1.0
geranial	6.0
methyl salicylate	m
citronellol	2.5
nerol	1.1
(E,Z)-2,6-dimethyl-3,5,7-octatrien-2-ol	*)
(E,E)-2,6-dimethyl-3,5,7-octatrien-2-ol	*)
geraniol	15.5
phenylethyl alcohol	m
phenylacetonitrile	m
β-ionone	m
(5R*,6R*)-3-oxo-7(E)-megastigmen-9-one	1.2
(3S*,5R*,6R*)-3-hydroxy-7(E)-megastigmen-9-one	m

Masdevallia striatella (cf. p. 100)

K 16740, BGSG, Nov. 20, 1990, 15.00–17.00, RK, c.

Compound	Area %
α-pinene	m
β-pinene	m
myrcene	1.1
limonene	m
(Z)-ocimene	1.5
(E)-ocimene	65.0
6-methyl-5-hepten-2-one	m
(Z)-3-hexenol	m
(E)-3(4)-epoxy-3,7-dimethyl-1,6-octadiene	m
(E,Z)-2,6-dimethyl-1,3,5,7-octatetraene	*)
(E,E)-2,6-dimethyl-1,3,5,7-octatetraene	*)
acetic acid	m
2,6,6-trimethyl-2-vinyl-tetrahydropyran-5-one	m
cis-linalool oxide (furanoid)	1.3
(E)-ocimene epoxide	1.5
benzaldehyde	m
linalool	15.8
caryophyllene	m
hotrienol	m
isovaleric acid	m
1,2-dimethoxy benzene	m
trans-linalool oxide (pyranoid)	m
cis-linalool oxide (pyranoid)	2.5
(E,Z)-2,6-dimethyl-3,5,7-octatrien-2-ol	*)
(E,E)-2,6-dimethyl-3,5,7-octatrien-2-ol	*)
(E)-geranylacetone	m
caproic acid	m
benzyl alcohol	m
vanilline	m
benzyl benzoate	m
benzyl salicylate	m

Masdevallia tridens (cf. p. 100)

K 16793, BGSG, Dec. 3, 1990, 15.00–17.00, RK, b.

Compound	Area %
isobutyl acetate	m
butyl acetate	1.2
isoamyl acetate	m
limonene	m
hexyl acetate	7.5
6-methyl-5-hepten-2-one	m
hexanol	m
heptyl acetate	m
3-octanol	m
trans-linalool oxide (furanoid)	m
octyl acetate	1.0
linalool	m
isocaryophyllene	m
caryophyllene	64.5
humulene	5.3
(E)-β-farnesene	m
decyl acetate	m
(Z)-4-decenyl acetate	1.1
benzyl acetate	1.0
(Z,E)-α-farnesene	m
(E,E)-α-farnesene	6.1

Maxillaria

Compound	Area %
(E,Z)-2,4-decadienal	1.3
benzyl butyrate	m
(E,E)-2,4-decadienal	m
phenylethyl acetate	m
geraniol	m
(E,Z)-2,4-decadienyl acetate	2.9
benzyl alcohol	m
(E,E)-2,4-decadienyl acetate	m
(E,Z)-2,4-decadienol	m
(E,E)-2,4-decadienol	m
geranyl butyrate	m
caryophyllene epoxide	m

Maxillaria nigrescens (cf. p. 106)

K 16798, BGSG, Dec. 3, 1990, 15.00–17.00, RK, b.

Compound	Area %
α-pinene	1.3
β-pinene	1.5
myrcene	m
butanol	m
limonene	3.2
eucalyptol	m
p-cymene	m
6-methyl-5-hepten-2-one	1.2
nonanal	1.4
decanal	m
α-copaene	3.2
(Z)-4-decenal	m
pentadecane	10.0
α-bergamotene	2.5
caryophyllene	2.3
methyl benzoate	1.6
cyclic β-ionone	2.0
(E)-β-farnesene	3.8
benzyl acetate	1.2
heptadecane	6.5
1-heptadecene	25.0
(Z)-4-decenol	1.5
phenylethyl acetate	1.0
dihydro-β-ionone	5.5
(E)-geranylacetone	2.6
phenylethyl alcohol	m
β-ionone	13.6

Maxillaria picta (cf. p. 106)

K 16745, BGSG, Nov. 20, 1990, 15.00–17.00, RK, c.

Compound	Area %
α-pinene	1.7
β-pinene	1.2
myrcene	m
limonene	3.6
(E)-ocimene	12.0
6-methyl-5-hepten-2-one	m
nonanal	1.0
(E,Z)-2,6-dimethyl-1,3,5,7-octatetraene	*)
(E,E)-2,6-dimethyl-1,3,5,7-octatetraene	*)
α-copaene	2.3
decanal	1.5
benzaldehyde	1.9
linalool	2.8
methyl benzoate	1.7
estragole	42.5
hydroquinone dimethyl ether	2.9
methyl salicylate	3.8
(E,Z)-2,6-dimethyl-3,5,7-octatrien-2-ol	*)
(E,E)-2,6-dimethyl-3,5,7-octatrien-2-ol	*)
benzyl 2-methylbutyrate	m
benzyl isovalerate	m
phenylethyl alcohol	2.8
methyl (E)-cinnamate	1.9
1.3,5-trimethoxy benzene	4.2
benzyl benzoate	m
benzyl salicylate	m

Maxillaria tenuifolia (cf. p. 106)

K 16160, BGSG, June 14, 1990, 13.00–14.30, RK, b.

Compound	Area %
α-pinene	1.5
myrcene	1.0
limonene	3.6
eucalyptol	4.2
6-methyl-5-hepten-2-one	2.5
nonanal	m
α-copaene	10.0
decanal	m
benzaldehyde	m
linalool	m
caryophyllene	5.2
2-undecanone	2.9
methyl benzoate	2.3
benzyl acetate	m
methyl salicylate	m

Compound	Area %
2-tridecanone	9.5
(E)-geranylacetone	2.3
2-pentadecanone	m
γ-decalactone	1.0
180(M⁺, 17), 137(55), 124(75), 81(19), 55(100)	3.5
δ-decalactone	29.5
212(M⁺?, 4), 124(17), 96(42), 82(83), 71(72), 41(100)	4.5

Maxillaria variabilis (yellow) (cf. p. 107)

K 17085, BGSG, March 11, 1991, 13.00–15.00, RK, b/c.

Compound	Area %
α-pinene	1.2
β-pinene	1.0
sabinene	m
myrcene	m
limonene	m
(E)-ocimene	3.0
6-methyl-5-hepten-2-one	m
(Z)-3-hexenol	m
nonanal	m
citronellal	3.1
α-copaene	2.0
decanal	m
(Z)-4-decenal	m
(E)-β-farnesene	5.3
neral	22.5
germacrene D	30.9
bicyclogermacrene	m
geranial	21.7
citronellol	1.9
nerol	1.2
geraniol	1.1
γ-decalactone	m

Maxillaria variabilis (dark red) (cf. p. 107)

K 17154, BGZH, April 2, 1991, 14.00–16.00, RK, b/c.

Compound	Area %
α-pinene	m
hexanal	m
β-pinene	m
sabinene	1.0
myrcene	1.4
heptan-2-one	4.0
limonene	3.0
eucalyptol	8.7
(E)-ocimene	1.0
octanal	m
2-heptanol	3.9
6-methyl-5-hepten-2-one	4.1
nonanal	2.3
α-copaene	10.9
decanal	3.8
benzaldehyde	m
linalool	m
caryophyllene	m
germacrene D	39.5
bicyclogermacrene	1.9
benzyl acetate	m
methyl salicylate	m
(E)-geranylacetone	3.9
benzyl alcohol	m
phenylethyl alcohol	m
(E)-nerolidol	m
γ-decalactone	m
benzyl benzoate	m

Miltonia regnellii (cf. p. 108)

K 16569, BGSG, Oct. 10, 1990, 15.00–17.00 RK, b.

Compound	Area %
α-pinene	1.0
β-pinene	m
hexanal	m
limonene	m
eucalyptol	m
(E)-ocimene	2.5
(Z)-4,8-dimethyl-1,3,7-nonatriene	1.2
(E)-4,8-dimethyl-1,3,7-nonatriene	75.3
6-methyl-5-hepten-2-one	m
hexanol	m
(Z)-3-hexenol	m
nonanal	m
decanal	m
benzaldehyde	m
linalool	m
methyl benzoate	m
dodecanol	m
(E)-2-dodecenal	11.2
phenylethyl alcohol	m
methyl (E)-cinnamate	1.0
methyl (E)-p-methoxycinnamate	m

Miltonia schroederiana (cf. p. 109)

K 16846, Weym., Dec. 8, 1990, 12.00–16.00, CW, b.

Compound	Area %
α-pinene	m
myrcene	m
limonene	m
(Z)-ocimene	1.5
(E)-ocimene	40.0
(E)-4,8-dimethyl-1,3,7-nonatriene	m
butyl caproate	m
hexyl butyrate	m
(E,Z)-2,6-dimethyl-1,3,5,7-octatetraene	*)
(E,E)-2,6-dimethyl-1,3,5,7-octatetraene	*)
trans-linalool oxide (furanoid)	m
α-cubebene	m
benzaldehyde	4.1
pentadecane	1.3
linalool	1.9
methyl benzoate	m
β-cubebene	2.1
benzyl acetate	1.0
methyl salicylate	m
(E,Z)-2,6-dimethyl-3,5,7-octatrien-2-ol	*)
(E,E)-2,6-dimethyl-3,5,7-octatrien-2-ol	*)
benzyl butyrate	m
benzyl alcohol	4,5
phenylethyl alcohol	m
3-phenylpropyl acetate	m
cinnamic aldehyde	m
3-phenylpropanol	m
phenylpropyl butyrate	m
(E)-cinnamyl acetate	9.0
eugenol	2.0
(E)-cinnamic alcohol	3.6
cinnamyl butyrate	1.0
chavicol	m
indole	m
benzyl benzoate	1.4
benzyl salicylate	m
methyl p-hydroxybenzoate	m

Miltonia spectabilis var. moreliana (cf. p. 109)

K 16503, BGHB, Sept. 19, 1990, 14.30–16.00, RK, a/b.

Compound	Area %
α-pinene	8.2
hexanal	m
β-pinene	1.0
sabinene	m
myrcene	4.3
limonene	3.5
eucalyptol	45.2
(E)-ocimene	13.0
(E,Z)-2,6-dimethyl-1,3,5,7-octatetraene	*)
(E,E)-2,6-dimethyl-1,3,5,7-octatetraene	*)
decanal	m
benzaldehyde	m
linalool	m
methyl benzoate	m
benzyl acetate	m
methyl salicylate	1.6
phenylethyl acetate	7.6
(E,Z)-2,6-dimethyl-3,5,7-octatrien-2-ol	*)
(E,E)-2,6-dimethyl-3,5,7-octatrien-2-ol	*)
(E)-geranylacetone	m
benzyl butyrate	m
benzyl alcohol	m
dodecyl acetate	m
(E)-nerolidol	3,8
methyl (E)-cinnamate	m
methyl 2-hydroxy-3-phenylpropionate	m

Miltoniopsis phalaenopsis (cf. p. 109)

K 16393, BGHB, Aug. 7/8, 1990, 23.00–8.00, RK, b.

Compound	Area %
limonene	2.8
octanal	m
6-methyl-5-hepten-2-one	m
nonanal	m
tetradecane	3.4
decanal	m
pentadecane	25.0
linalool	1.0
caryophyllene	3.2
geranyl acetate	m
benzyl acetate	4.5
(E)-geranylacetone	m
benzyl alcohol	m
M+(?), 204(3), 165(5), 109(15), 69(62), 41(100)	5.5
γ-decalactone	m
(E)-2(3)-dihydrofarnesyl acetate	18.4
elemicine	17.8
(E)-2(3)-dihydrofarnesol	4.0
(E)-isoelemicine	1.0
benzyl benzoate	5.0

Neofinetia falcata (cf. p. 167)

K 17255, BGZH, April 30/May 1, 1991, 20.00–7.00, RK, a.

Compound	Area %
methyl 2-methylbutyrate	m
α-pinene	m
β-pinene	m
butanol	m
myrcene	m
methyl tiglate	4.5
limonene	1.0
2-methylbutanol	m
octanal	m
(Z)-3-hexenyl acetate	m
6-methyl-5-hepten-2-one	m
(Z)-3-hexenol	m
nonanal	m
butyl tiglate	m
isomenthone	m
(Z)-3-hexenyl butyrate	m
isoamyl tiglate	m
(Z)-3-hexenyl 2-methylbutyrate	m
decanal	m
benzaldehyde	m
2-methylbutyl tiglate	m
pentadecane	1.0
linalool	30.5
(E)-2-methyl-2-buten-1-yl tiglate	m
methyl benzoate	16.1
hexadecane	m
hexyl tiglate	m
isomenthol	m
ethyl benzoate	m
(Z)-3-hexenyl angelate	m
(Z)-3-hexenyl tiglate	21.5
benzyl acetate	m
(Z)-3-hexenyl (Z)-3-hexenoate	m
(Z,E)-α-farnesene	m
(E,E)-α-farnesene	1.0
methyl salicylate	m
(E,E)-4,8,12-trimethyl-1,3,7,11-tridecatetraene	m
(E)-geranylacetone	m
benzyl 2-methylbutyrate	m
benzyl alcohol	7.5
isoamyl benzoate	m
202(M+, 10), 159(38), 117(24), 91(52), 67(88), 41(100)	m
(E)-cinnamic aldehyde	m
(E)-2-methyl-2-buten-1-yl benzoate	m
hexyl benzoate	m
benzyl tiglate	2.0
(Z)-3-hexenyl benzoate	m
γ-decalactone	m
δ-decalactone	m
jasmine lactone	m
220(M+, 1), 162(16), 159(12), 105(21), 93(29), 41(100)	1.1
(E,E)-farnesol	m
indole	m
vanilline	m
benzyl benzoate	m

Nigritella nigra (cf. p. 176)

K 16313, Gr. St. Bernhard/Switzerland, July 28, 1990, 13.00–16.00, RK, a/b.

Compound	Area %
isovaleraldehyde	1.0
methyl isovalerate	4.0
α-pinene	m
β-pinene	m
myrcene	m
heptanal	m
limonene	8.0
eucalyptol	1.5
isoamyl alcohol	4.0
ethyl caproate	m
isoamyl isovalerate	m
(Z)-3-hexenyl acetate	m
6-methyl-5-hepten-2-one	m
hexanol	m
(Z)-3-hexenol	m
nonanal	m
p-methyl anisole	m
1-octen-3-ol	m
heptanol	m
methyl 2-hydroxy-3-methylvalerate	m
methyl 2-hydroxy-4-methylvalerate	m
benzaldehyde	1.5
linalool	1.5
octanol	m
methyl benzoate	2.2
phenylacetaldehyde	1.0
nonanol	m
benzyl formate	m
α-terpineol	m
benzyl acetate	m
indole	m
phenylethyl formate	1.0
damascenone	m
phenylethyl acetate	m
(E)-geranylacetone	m
benzyl alcohol	10.5
benzyl isovalerate	m

Odontoglossum

phenylethyl alcohol	47.0
phenylethyl isovalerate	m
p-cresol	m
6,10,14-trimethyl-pentadecan-2-one	m
eugenol	m
vanillyl ethyl ether	m
M+(?), 138(50), 124(27), 120(45), 109(49), 69(100)	m
vanilline	m
benzyl benzoate	m
phenylethyl benzoate	m

Odontoglossum cirrhosum (cf. p. 110)

K 16519, BGHB, Nov. 19, 1990, 12.00–14.30, RK, b.

Compound	Area %
α-pinene	m
hexanal	m
myrcene	1.0
limonene	1.0
(E)-ocimene	35.0
hexanol	m
(E,Z)-2,6-dimethyl-1,3,5,7-octatetraene	*)
(E,E)-2,6-dimethyl-1,3,5,7-octatetraene	*)
(Z)-5-octenyl acetate	m
trans-linalool oxide (furanoid)	m
3-methyl-2-cyclopenten-1-one	m
benzaldehyde	8.0
linalool	32.2
(E,Z)-2,6-dimethyl-3,5,7-octatrien-2-ol	*)
(E,E)-2,6-dimethyl-3,5,7-octatrien-2-ol	*)
benzyl alcohol	5.0
phenylethyl alcohol	4.5
methyl (E)-cinnamate	1.1
2-amino benzaldehyde	4.2
benzyl benzoate	m

Odontoglossum constrictum (cf. p. 112)

K 13225, BGZH, Nov. 21, 1987, 9.00–16.00, RK, a/b.

Compound	Area %
hexanal	m
myrcene	m
limonene	m
(Z)-ocimene	m
(E)-ocimene	28.0
hexyl acetate	m
(E)-4,8-dimethyl-1,3,7-nonatriene	m
(Z)-3-hexenyl acetate	m
anisole	m
hexanol	m
(Z)-3-hexenol	m
methyl octanoate	3.2
(E,Z)-2,6-dimethyl-1,3,5,7-octatetraene	*)
(E,E)-2,6-dimethyl-1,3,5,7-octatetraene	*)
ethyl octanoate	m
benzaldehyde	7.8
methyl nonanoate	m
linalool	1.0
caryophyllene	6.9
phenylacetaldehyde	m
methyl decanoate	1.8
humulene	m
benzyl acetate	1.1
1,2-dimethoxy benzene	4.1
3-phenylpropanal	1.8
phenylethyl acetate	5.5
(E,Z)-2,6-dimethyl-3,5,7-octatrien-2-ol	*)
(E,E)-2,6-dimethyl-3,5,7-octatrien-2-ol	*)
benzyl alcohol	9.1
phenylethyl alcohol	6.0
3-phenylpropyl acetate	3.2
anis aldehyde	1.6
(Z)-cinnamyl acetate	m
(E)-cinnamic aldehyde	1.1
caryophyllene epoxide	m
3-phenylpropanol	1.0
1-nitro-2-phenylethane	m
(E)-cinnamyl acetate	1.8
2-amino benzaldehyde	11.1
anisyl acetate	m
(E)-cinnamic alcohol	m
benzyl benzoate	m

Odontoglossum pendulum (cf. p. 113)

K 17470, BGZH, June 17, 1991, 12.00–14.00, RK, b/c.

Compound	Area %
limonene	1.0
hexyl acetate	1.3
(Z)-3-hexenyl acetate	m
6-methyl-5-hepten-2-one	m
hexanol	2.7
(Z)-3-hexenol	1.3
nonanal	m
6-methyl-5-hepten-2-yl acetate	1.0
6-methyl-5-hepten-2-ol	m
octyl acetate	m
decanal	m
benzaldehyde	m
(E)-2-nonenal	1.0
nonyl acetate	m

Compound	Area %
(Z)-3-nonenyl acetate	1.1
methyl benzoate	m
dodecanal	m
hydroquinone dimethyl ether	8.7
methyl salicylate	5.6
(E)-geranylacetone	3.8
benzyl alcohol	5.8
6,10-dimethyl-5,9-undecadien-2-yl acetate	18.5
nonadecane	7.0
anis aldehyde	1.0
methyl myristate	1.0
(E)-cinnamic aldehyde	3.2
6,10-dimethyl-5,9-undecadien-2-ol	m
methyl pentadecanoate	m
(E)-cinnamyl acetate	m
methyl palmitate	17.2
methyl anthranilate	m
anisyl alcohol	m
(E)-cinnamic alcohol	5.8
benzyl benzoate	m

Odontoglossum pulchellum (cf. p. 113)

K 17119, BGSG, March 17, 1991, 11.00–12.45, RK, a.

Compound	Area %
limonene	m
prenol	m
hexanol	m
benzyl methyl ether	1.6
(Z)-3-hexenol	m
p-methyl anisole	m
benzaldehyde	1.0
linalool	m
caryophyllene	m
methyl benzoate	m
neral	1.0
methyl geranate	m
methyl decanoate	m
benzyl acetate	m
geranial	7.8
hydroquinone dimethyl ether	25.5
methyl salicylate	1.9
geranyl acetate	m
citronellol	m
p-methoxybenzyl methyl ether	m
nerol	1.0
geraniol	24.8
benzyl alcohol	25.9
phenylethyl alcohol	m
anis aldehyde	m
methyl eugenol	1.5
methyl 2-methoxybenzoate	m
1,2,4-trimethoxy benzene	m
anisyl acetate	1.1
eugenol	m
anisyl alcohol	m
p-methoxyphenylethyl alcohol	m
benzyl benzoate	m

Oncidium longipes (cf. p. 115)

K 17162, BGZH, April 2, 1991, 14.30–16.00, RK, b/c.

Compound	Area %
α-pinene	m
hexanal	2.8
myrcene	2.6
limonene	2.8
(E)-ocimene	m
(E)-4,8-dimethyl-1,3,7-nonatriene	2.0
6-methyl-5-hepten-2-one	2.6
(Z)-3-hexenol	m
nonanal	m
1-octen-3-ol	2.7
decanal	1.2
benzaldehyde	8.5
linalool	2.5
methyl benzoate	m
hotrienol	m
methyl phenylacetate	m
methyl salicylate	1.7
(E,E)-4,8,12-trimethyl-1,3,7,11-tridecatetraene	1.0
(E)-geranylacetone	1.2
benzyl alcohol	40.5
M$^+$(?), 126(3), 111(9), 82(100), 54(37), 39(33)	m
anis aldehyde	3.0
(E)-nerolidol	4.9
methyl anisate	m
anisyl alcohol	4.7
(E)-cinnamic alcohol	8.0
benzyl benzoate	m

Oncidium ornithorhynchum (cf. p. 118)

K 17471, BGZH, July 3, 1991, 9.00–12.00, RK, a/b.

Compound	Area %
α-pinene	7.5
hexanal	m
(E)-2-methyl-2-butenal	m
β-pinene	1.2
sabinene	1.3
myrcene	8.0

Oncidium, Peristeria

limonene	1.3
eucalyptol	4.1
3-octanone	m
acetoin	m
octanal	m
(E)-2-methyl-2-buten-1-ol	m
6-methyl-5-hepten-2-one	2.5
chalcogran A	m
chalcogran B	m
hexanol	m
(Z)-3-hexenol	m
nonanal	m
trans-sabinene hydrate	1.7
trans-linalool oxide (furanoid)	2.3
decanal	m
benzaldehyde	m
linalool	6.0
methyl benzoate	m
(E)-2-nonenal	m
neral	4.2
(Z)-3-nonen-1-ol	m
methyl geranate	m
α-terpineol	9.8
geranial	18.0
citronellol	m
nerol	8.7
(E,E)-4,8,12-trimethyl-1,3,7,11-tridecatetraene	m
geraniol	7.8
benzyl alcohol	m
β-ionone	m
hexadecanal	3.5
2-amino benzaldehyde	m
ethyl palmitate	1.7
vanilline	m

Oncidium sarcodes (cf. p. 115)

K 16256, BGSG, July 9, 1990, 12.00–14.15, RK, c.

Compound	Area %
α-pinene	1.1
limonene	1.5
(Z)-3-hexenal	1.0
(Z)-ocimene	m
(E)-ocimene	18.0
6-methyl-5-hepten-2-one	m
hexanol	1.0
(Z)-3-hexenol	6.3
nonanal	1.2
(E,E)-2,6-dimethyl-1,3,5,7-octatetraene	*)
cis-linalool oxide (furanoid)	1.4
decanal	m
benzaldehyde	1.2
linalool	13.0
acetophenone	m
decyl acetate	m
benzyl acetate	14.0
(E,E)-α-farnesene	2.0
methyl salicylate	1.8
phenylethyl acetate	5.5
(E,E)-2,6-dimethyl-3,5,7-octatrien-2-ol	*)
benzyl alcohol	1.2
dodecyl acetate	2.4
phenylethyl alcohol	m
tetradecanal	2.4
anis aldehyde	m
methyl eugenol	m
tetradecyl acetate	2.9
6,10,14-trimethyl-pentadecan-2-one	2.8

Oncidium tigrinum (cf. p. 115)

K 13293, BGSG, Dec. 5, 1987, 11.00–14.00, RK, a.

Compound	Area %
myrcene	m
limonene	m
eucalyptol	3.8
(Z)-ocimene	m
(E)-ocimene	67.0
6-methyl-5-hepten-2-one	m
hexanol	m
(Z)-3-hexenol	m
methyl octanoate	m
(E,Z)-2,6-dimethyl-1,3,5,7-octatetraene	*)
(E,E)-2,6-dimethyl-1,3,5,7-octatetraene	*)
benzaldehyde	m
linalool	m
methyl benzoate	4.9
ethyl benzoate	m
cyclic β-ionone	0.5
1,2-dimethoxy benzene	m
phenylethyl acetate	m
(E,Z)-2,6-dimethyl-3,5,7-octatrien-2-ol	*)
dihydro-β-ionone	1.4
(E,E)-2,6-dimethyl-3,5,7-octatrien-2-ol	*)
β-ionone	14.5

Peristeria elata (cf. p. 118)

K 16505, BGHB, July 19, 1990, 11.30–14.00, RK, a.

Compound	Area %
α-pinene	m
β-pinene	m

isoamyl acetate	m
myrcene	4.0
limonene	3.9
eucalyptol	48.5
prenyl acetate	m
p-cymene	m
terpinen-4-ol	m
α-terpineol	m
benzyl acetate	m
phenylethyl acetate	40.6
phenylethyl alcohol	m
benzyl benzoate	m

Pescatorea cerina (cf. p. 57)

K 16346, BGHB, Aug. 8, 1990, 10.30–12.30, RK, b/c.

Compound	Area %
α-pinene	3.0
β-pinene	1.0
myrcene	m
heptanal	m
limonene	4.5
octanal	m
6-methyl-5-hepten-2-one	1.0
nonanal	1.0
3-octanol	1.5
decanal	2.4
benzaldehyde	2.0
(Z)-4-decenal	13.6
linalool	8.0
methyl benzoate	m
ethyl benzoate	m
α-terpineol	m
benzyl acetate	1.5
(E,Z)-2,4-decadienal	3.6
methyl salicylate	14.0
ethyl salicylate	m
(E,E)-2,4-decadienal	1.6
(Z)-4-decenol	2.2
phenylethyl acetate	m
tridecanal	1.0
(E)-geranylacetone	1.6
benzyl alcohol	1.4
(Z,Z)-4,7-tridecadienal (tentative)	6.0
elemicine	m
6-methoxyeugenol	m
methyl (Z)-p-methoxycinnamate	m
methyl (E)-p-methoxycinnamate	m
benzyl benzoate	15.0

Pescatorea dayana (cf. p. 57)

K 16254, BGSG, July 9, 1990, 12.00–14.15, RK, b.

Compound	Area %
α-pinene	1.0
myrcene	1.0
limonene	1.8
eucalyptol	1.3
6-methyl-5-hepten-2-one	m
nonanal	m
cis-linalool oxide (furanoid)	1.3
decanal	m
benzaldehyde	1.0
linalool	3.5
phenylacetaldehyde	m
methyl benzoate	m
neral	m
methyl geranate	1.0
α-terpineol	m
benzyl acetate	21.9
geranial	1.5
geranyl acetate	1.6
nerol	1.7
phenylethyl acetate	5.3
geraniol	33.0
benzyl alcohol	m
phenylethyl alcohol	1.4
methyl eugenol	1.4
(E)-cinnamic aldehyde	m
methyl (E,E)-3,7,11-trimethyl-2,6,10-dodecatrienoate	m
elemicine	m
(E,E)-farnesyl acetate	5.2
(E)-cinnamic alcohol	m
(E,E)-farnesol	1.4
indole	m
benzyl benzoate	6.1

Phalaenopsis violacea Borneo type (cf. p. 169)

K 16494, BGHB, Sept. 19, 1990, 11.00–14.00, RK, a/b.

Compound	Area %
myrcene	m
limonene	m
6-methyl-5-hepten-2-one	m
(Z)-3-hexenol	m
nonanal	m
trans-linalool oxide (furanoid)	m
linalool	49.0

Compound	Area %
neral	m
citronellol	1.0
nerol	m
geraniol	43.0
benzyl alcohol	m

Phalaenopsis violacea Malaya type
(cf. p. 169)

K 15401, BGSG, Oct. 27, 1989, 12.30–14.00, RK, b.

Compound	Area %
α-pinene	m
α-phellandrene	m
myrcene	m
limonene	m
(Z)-3-hexenyl acetate	m
6-methyl-5-hepten-2-one	m
(Z)-3-hexenol	1.0
nonanal	m
trans-linalool oxide (furanoid)	m
decanal	m
benzaldehyde	m
linalool	27.8
caryophyllene	1.1
terpinen-4-ol	m
methyl benzoate	m
humulene	m
neral	m
germacrene D	1.9
benzyl acetate	m
geranial	m
methyl salicylate	1.8
geranyl acetate	1.0
trans-linalool oxide (pyranoid)	m
nerol	m
phenylethyl acetate	m
(E,E)-4,8,12-trimethyl-1,3,7,11-tridecatetraene	m
geraniol	8.3
benzyl alcohol	1.0
phenylethyl alcohol	m
3-phenylpropyl acetate	m
methyl eugenol	m
(E)-cinnamic aldehyde	3.6
3-phenylpropanol	m
(E)-nerolidol	m
methyl (E)-cinnamate	m
(Z)-3-hexenyl benzoate	m
(E)-cinnamyl acetate	3.7
eugenol	m
(Z)-cinnamic alcohol	m
elemicine	26.7
(E)-cinnamic alcohol	9.2
benzyl benzoate	m

Platanthera bifolia (cf. p. 177)

K 17446, Hostig/CH-Uster, June 23, 1991, 19.30–23.30, RK, a.

Compound	Area %
α-pinene	m
myrcene	m
limonene	m
(E)-ocimene	7.0
(Z)-3-hexenyl acetate	m
6-methyl-5-hepten-2-one	m
hexanol	m
(E,Z)-2,6-dimethyl-1,3,5,7-octatetraene	*)
(E,E)-2,6-dimethyl-1,3,5,7-octatetraene	*)
trans-2-acetyl-5-vinyl-5-methyl-THF	m
cis-2-acetyl-5-vinyl-5-methyl-THF	m
benzaldehyde	m
lilac aldehyde a	m
lilac aldehyde b	m
linalool	18.8
lilac aldehyde c	1.0
lilac aldehyde d	m
caryophyllene	1.0
methyl benzoate	58.0
neral	m
lilac alcohol b	m
benzyl acetate	m
geranial	m
acetate of lilac alcohol a	m
lilac alcohol d	m
acetate of lilac alcohol c	m
methyl salicylate	2.4
lilac alcohol a	m
nerol	m
(E,Z)-2,6-dimethyl-3,5,7-octatrien-2-ol	*)
(E,E)-2,6-dimethyl-3,5,7-octatrien-2-ol	*)
lilac alcohol c	m
p-cymen-8-ol	m
geraniol	m
benzyl alcohol	1.0
benzyl isovalerate	m
3-phenylpropyl acetate	m
(E)-cinnamic aldehyde	m
3-phenylpropanol	m
(E)-cinnamyl acetate	m
eugenol	m
(E)-cinnamic alcohol	m
(E)-isoeugenol	m
vanilline	m

Platanthera chlorantha (cf. p. 178)

K 17447, Hostig/CH-Uster, June 25/26, 1991, 20.30–3.00, RK, a.

Compound	Area %
benzyl benzoate	1.8
benzyl salicylate	m
α-pinene	m
sabinene	m
myrcene	m
limonene	m
(Z)-ocimene	m
(E)-ocimene	4.0
6-methyl-5-hepten-2-one	m
(E,Z)-2,6-dimethyl-1,3,5,7-octatetraene	*)
(E,E)-2,6-dimethyl-1,3,5,7-octatetraene	*)
benzaldehyde	m
linalool	45.0
caryophyllene	m
methyl benzoate	30.0
neral	m
geranial	m
methyl salicylate	1.2
nerol	m
(E,Z)-2,6-dimethyl-3,5,7-octatrien-2-ol	*)
(E,E)-2,6-dimethyl-3,5,7-octatrien-2-ol	*)
geraniol	15.3
benzyl alcohol	m
benzyl benzoate	1.0
benzyl salicylate	m

Plectrelminthus caudatus (cf. p. 138)

K 16537, BGSG, Oct. 1, 1990, 19.00–21.00, RK, a.

Compound	Area %
methyl butyrate	m
methyl 2-methylbutyrate	m
α-pinene	m
hexanal	m
β-pinene	m
sabinene	m
myrcene	m
methyl tiglate	1.5
methyl caproate	3.0
butyl butyrate	m
butyl 2-methylbutyrate	m
2-methylbutyl butyrate	m
2-methylbutyl 2-methylbutyrate	m
methyl heptanoate	m
(E)-2-methyl-2-buten-1-yl propionate	m
propyl tiglate	m
hexanol	m
(E)-2-methyl-2-buten-1-yl butyrate	1.0
(E)-2-methyl-2-buten-1-yl 2-methylbutyrate	1.0
hexyl butyrate	3.4
butyl tiglate	5.0
hexyl 2-methylbutyrate	1.7
hexyl valerate	m
2-methylbutyl tiglate	1.8
amyl tiglate	1.6
linalool	1.5
(E)-2-methyl-2-buten-1-yl caproate	m
caryophyllene	m
(E)-2-methyl-2-buten-1-yl tiglate (94)	6.9
hexyl tiglate	58.0
humulene	m
heptyl tiglate	m
(E,E)-α-farnesene	2.5
(E)-2-methyl-2-buten-1-yl benzoate	m
(E)-cinnamic aldehyde	m
(E)-2(3)-epoxy-2-methylbutyl tiglate	m
methyl linoleate	m
benzyl benzoate	m

(E)-2-Methyl-2-buten-1-yl tiglate (94)

Synthesis:
Esterification of (E)-2-methyl-2-buten-1-ol with tiglic acid chloride [71].

Spectral data:
MS: 168(M$^+$, 2), 153(1), 139 (9), 123(8), 101(13), 83(100), 69(32), 68(16), 55(56), 41(43).

Polystachya campyloglossa (cf. p. 139)

K 16542, BGSG, Oct. 1, 1990, 17.00–19.00, RK, b.

Compound	Area %
α-pinene	10.0
butyl acetate	m
β-pinene	2.2
sabinene	1.0
heptanal	m
limonene	3.5
isoprenyl acetate	42.0
isoprenol	16.0
prenyl acetate	12.0
prenol	m

Polystachya

Compound	Area %
6-methyl-5-hepten-2-one	1.4
(Z)-3-hexenol	m
nonanal	m
p-methyl anisole	m
decanal	m
benzaldehyde	m
linalool	1.2
methyl benzoate	m
benzyl acetate	m
methyl salicylate	m
dihydro-β-ionone	m
(E)-geranylacetone	m
benzyl alcohol	m
phenylethyl alcohol	1.0
β-ionone	m
prenyl benzoate	m
benzyl benzoate	m
phenylethyl benzoate	m
benzyl salicylate	m

Polystachya cultriformis (cf. p. 141)

K 16368, BGSG, Aug. 14, 1990, 12.00–14.00, RK, b.

Compound	Area %
myrcene	m
limonene	1.8
(E)-ocimene	m
octanal	m
(E)-4,8-dimethyl-1,3,7-nonatriene	m
6-methyl-5-hepten-2-one	4.5
hexanol	m
(Z)-3-hexenol	m
nonanal	m
decanal	m
benzaldehyde	m
hydroquinone dimethyl ether	m
methyl salicylate	m
citronellol	m
dihydro-β-ionone	m
(E)-geranylacetone	1.6
benzyl butyrate	2.5
benzyl alcohol	8.5
phenylethyl alcohol	m
β-ionone	m
2-N-methylamino benzaldehyde	8.5
germacra-1(10).5-dien-4-ol	7.5
(E)-nerolidol	39.5
benzyl caproate	1.1
γ-decalactone	m
2-amino benzaldehyde	m
(E)-2(3)-dihydrofarnesol	4.4
(Z,E)-farnesol	m
(E,E)-farnesol	3.9
methyl *cis*-(Z)-jasmonate	m
benzyl benzoate	2.6

Polystachya fallax (cf. p. 141)

K 16371, BGSG, Aug. 14, 1990, 12.00–14.00, RK, b.

Compound	Area %
methyl butyrate	m
methyl 2-methylbutyrate	5.0
α-pinene	1.0
α-thujene	1.2
methyl valerate	m
β-pinene	m
sabinene	27.0
myrcene	3.0
methyl caproate	m
limonene	4.5
γ-terpinene	3.0
p-cymene	1.5
terpinolene	2.5
octanal	m
6-methyl-5-hepten-2-one	1.2
methyl 2-oxo-3-methylvalerate	1.2
nonanal	1.3
methyl 2-hydroxyvalerate	m
trans-sabinene hydrate	2.8
citronellal	m
methyl 2-hydroxy-3-methylvalerate	2.8
decanal	1.2
methyl 2-hydroxy-3-methylvalerate	m
methyl 2-hydroxy-4-methylvalerate	m
linalool	1.0
ethyl 3,7-dimethyl-6-octenoate	m
undecanal	m
caryophyllene	m
terpinen-4-ol	m
methyl benzoate	6.0
α-terpineol	m
(E,E)-α-farnesene	1.9
citronellol	m
citronellyl butyrate	m
(E)-geranylacetone	m
benzyl alcohol	m
citronellyl valerate	8.0
β-ionone	m
citronellyl caproate	m
M+(?), 176(32), 161(68), 119(65), 105(90), 41(100)	m
methyl 2-methoxybenzoate	1.5
methyl (E)-cinnamate	m
methyl anthranilate	1.5
spathulenol	m

elemicine	m
methyl (Z)-p-methoxycinnamate	1.5
indole	5.5
methyl (E)-p-methoxycinnamate	1.8
benzyl benzoate	m

Polystachya mazumbaiensis (cf. p. 142)

K 16370, BGSG, Oct. 1, 1990, 12.00–14.00, RK, c.

Compound	Area %
methyl 2-methylbutyrate	m
α-pinene	m
ethyl butyrate	m
hexanal	2.5
myrcene	m
limonene	4.2
isoamyl alcohol	m
octanal	m
hexyl acetate	m
6-methyl-5-hepten-2-one	1.3
hexanol	m
nonanal	m
p-methyl anisole	m
decanal	1.0
pentadecane	1.5
benzaldehyde	1.0
hexadecane	m
methyl (Z)-4-decenoate	m
(Z)-3-nonen-1-ol	m
heptadecane	2.0
1-heptadecene	1.3
benzyl acetate	m
hydroquinone dimethyl ether	m
2-tridecanone	1.8
dihydro-β-ionone	1.5
(E)-geranylacetone	45.0
nonadecane	2.3
phenylethyl alcohol	2.7
β-ionone	m
jasmone	6.4
2-N-methylamino benzaldehyde	5.7
M+(?), 176(32), 161(68), 119(65), 105(90), 41(100)	2.0
(E)-cinnamic aldehyde	1.0
eicosane	1.0
anisyl alcohol	1.5
anisyl butyrate	1.0
anisyl caproate	1.0
benzyl benzoate	m

Rangaeris amaniensis (cf. p. 138)

K 16339, BGHB, Aug. 7/8 1990, 23.00–8.00, RK, a/b.

Compound	Area %
α-pinene	m
hexanal	m
limonene	1.0
6-methyl-5-hepten-2-one	m
nonanal	m
benzaldehyde	6.5
linalool	2.4
methyl benzoate	2.4
benzyl acetate	m
δ-cadinene	1.7
methyl salicylate	5.8
phenylethyl acetate	m
geraniol	3.6
phenylethyl alcohol	8.5
methyl eugenol	m
methyl (E)-cinnamate	20.1
methyl anthranilate	1.0
(E,E)-farnesal	2.4
(E,E)-farnesol	27.9
indole	3.0
vanilline	m

Rhynchostylis coelestis (cf. p. 171)

K 16148, BGSG, June 14, 1990, 12.30–14.30, RK, a/b.

Compound	Area %
myrcene	m
limonene	m
eucalyptol	1.1
(Z)-ocimene	1.1
(E)-ocimene	47.0
p-cymene	m
6-methyl-5-hepten-2-one	m
hexanol	m
(E,Z)-2,6-dimethyl-1,3,5,7-octatetraene	*)
(E,E)-2,6-dimethyl-1,3,5,7-octatetraene	*)
pentadecane	m
linalool	m
α-terpineol	m
benzyl acetate	m
geranial	m
(E,Z)-2,6-dimethyl-3,5,7-octatrien-2-ol	*)
(E,E)-2,6-dimethyl-3,5,7-octatrien-2-ol	*)
(E)-geranylacetone	m
anethole	m
(Z,E)-farnesal	5.9

Rodriguezia, Stanhopea, Trichocentrum

(E,E)-farnesal	34.4
(Z,E)-farnesol	m
(E,E)-farnesol	3.8

Rodriguezia refracta (cf. p. 119)

K 17433, Weym., June 2, 1991, 9.00–15.00, CW, a.

Compound	Area %
limonene	m
linalool	m
(E,Z)-2,4-decadienal	89.0
(E,E)-2,4-decadienal	3.8
(E,Z)-2,4-decadienyl acetate	3.5
(E,E)-2,4-decadienyl acetate	m
(E,Z)-2,4-decadienol	2.0

Stanhopea jenischiana (cf. p. 79)

K 17148, BGZH, April 2, 1991, 15.00–16.30, RK, a.

Compound	Area %
styrene	4.2
benzaldehyde	m
methyl benzoate	m
methyl (Z)-cinnamate	1.0
methyl (E)-cinnamate	93.2

Stanhopea oculata (cf. p. 80)

K 15290, BGSG, Sept. 25, 1989, 11.30–13.45, RK, a.

Compound	Area %
α-pinene	2.0
α-thujene	m
β-pinene	1.0
sabinene	m
myrcene	5.4
limonene	1.5
eucalyptol	86.0
p-cymene	m
trans-sabinene hydrate	m
cis-sabinene hydrate	m
linalool	m
terpinen-4-ol	m
δ-terpineol	m
α-terpineol	1.0
phenylethyl acetate	m
phenylethyl alcohol	m

Stanhopea tigrina (cf. p. 80)

K 16252, BGSG, July 9, 1990, 12.00–14.15, RK, a.

Compound	Area %
limonene	m
eucalyptol	m
nonanal	m
acetic acid	m
benzaldehyde	m
linalool	m
methyl benzoate	1.0
benzyl acetate	1.4
methyl salicylate	1.6
phenylethyl acetate	92.0
dihydro-β-ionone	m
benzyl alcohol	m
phenylethyl alcohol	1.0
β-ionone	m
coumarine	m
vanilline	m
benzyl benzoate	m
p-hydroxy phenylbutan-2-one	m

Trichocentrum tigrinum (cf. p. 119)

K 17560, Weym., July 14, 1991, 9.30–15.00, CW, b.

Compound	Area %
myrcene	m
limonene	6.0
eucalyptol	m
nonanal	m
decanal	m
benzaldehyde	m
linalool	2.4
methyl benzoate	18.0
benzyl acetate	11.4
(E)-geranylacetone	m
(E)-nerolidol	m
(E)-2(3)-dihydrofarnesyl acetate	1.0
(Z,E)-farnesal	6.8
(Z,E)-farnesyl acetate	10,0
(E,E)-farnesal	7.5
(E,E)-farnesyl acetate	2.3
(E)-2(3)-dihydrofarnesol	3.0
220(M$^+$, 1), 162(16), 159(12), 105(21), 93(29), 41(100)	4.0
(Z,E)-farnesol	2.3
(E,E)-farnesol	5.0
benzyl benzoate	m

Trichoglottis philippinensis (cf. p. 171)

K 16481, BGSG, Sept. 14, 1990, 12.30–14.30, RK, b/c.

Compound	Area %
6-methyl-5-hepten-2-one	1.0
p-methyl anisole	7.2
decanal	m
benzaldehyde	5.0
linalool	19.2
(Z,E)-α-farnesene	1.5
(E,E)-α-farnesene	40.5
methyl salicylate	m
3-phenylpropanal	m
benzyl alcohol	3.0
phenylethyl alcohol	m
3-phenylpropanol	m
220(M+, 1), 162(12), 159(10), 93(25), 69(20), 43(100)	11.5
benzyl salicylate	m

Trixspermum arachnites (cf. p. 172)

K 16930, Weym., Jan. 12, 1991, 12.00–17.00, CW, a.

Compound	Area %
hexanal	m
sabinene	m
myrcene	m
ethyl caproate	m
acetoin	m
octanal	m
(E)-4,8-dimethyl-1,3,7-nonatriene	m
ethyl octanoate	m
trans-linalool oxide (furanoid)	1.4
cis-linalool oxide (furanoid)	46.0
decanal	m
linalool	1.6
ethyl decanoate	1.0
ethyl (Z)-4-decenoate	2.4
(E,Z)-2,4-decadienal	m
cis-linalool oxide (pyranoid)	m
methyl laurate	m
ethyl (E,Z)-2,4-decadienoate	1.7
ethyl laurate	2.5
ethyl (E,E)-2,4-decadienoate	m
(E)-nerolidol	37.0
ethyl tetradecanoate	m
(E)-2(3)-dihydrofarnesol	1,9
δ-decalactone	m

Vanda coerulescens (cf. p. 173)

K 17206, Kais., April 13, 1991, 9.00–15.00, RK, b.

Compound	Area %
α-pinene	1.8
hexanal	1.9
β-pinene	1.2
limonene	5.5
isoamyl alcohol	m
(Z)-3-hexenal	0.4
octanal	m
6-methyl-5-hepten-2-one	m
(Z)-3-hexenol	0.3
nonanal	m
tetradecane	2.2
α-cubebene	m
decanal	m
pentadecane	5.1
hexadecane	4.5
benzyl acetate	m
δ-cadinene	4.4
methyl (E,Z)-2,4-decadienoate	6.0
methyl (x,y,z)-decatrienoate	27.0
methyl (E,E)-2,4-decadienoate	m
phenylethyl alcohol	m
2-amino benzaldehyde	1.8
methyl anthranilate	23.5
methyl N-formyl-anthranilate	m
benzyl benzoate	m

Vanda denisoniana (cf. p. 175)

K 15955, Weym., Febr. 5, 1990, 10.00–14.00, CW, b.

Compound	Area %
α-pinene	m
β-pinene	m
myrcene	m
limonene	m
eucalyptol	m
(E)-ocimene	14.0
(E,Z)-2,6-dimethyl-1,3,5,7-octatetraene	*)
(E,E)-2,6-dimethyl-1,3,5,7-octatetraene	*)
trans-sabinene hydrate	m
(E)-ocimene epoxide	m
benzaldehyde	m
linalool	66.0
methyl benzoate	m
(E)-β-farnesene	m
benzyl acetate	m
(Z,E)-α-farnesene	2.5
(E,E)-α-farnesene	m

Compound	Area %
phenylethyl acetate	m
(E,Z)-2,6-dimethyl-3,5,7-octatrien-2-ol	*)
(E,E)-2,6-dimethyl-3,5,7-octatrien-2-ol	*)
benzyl alcohol	5.6
phenylethyl alcohol	m
phenylacetonitrile	1.0
2-amino benzaldehyde	m
methyl anthranilate	1.5
220(M^+, 1), 162(16), 159(12), 105(21), 93(29), 41(100)	1.0

Vanda tessellata (cf. p. 175)

K 17109, BGZH, March 15, 1991, 11.00–14.00, RK, a.

Compound	Area %
methyl isobutyrate	m
methyl 2-methylbutyrate	m
α-pinene	m
myrcene	m
(Z)-ocimene	m
(E)-ocimene	m
benzaldehyde	1.1
linalool	23.0
methyl benzoate	61.5
ethyl benzoate	m
benzyl acetate	m
methyl salicylate	m
3-phenylpropanal	m
α-ionone	m
(E)-geranylacetone	m
benzyl alcohol	m
(Z)-cinnamic aldehyde	m
methyl (Z)-cinnamate	m
(E)-cinnamic aldehyde	5.1
3-phenylpropanol	m
methyl (E)-cinnamate	4.6
p-cresol	m
(E)-cinnamic alcohol	m
indole	m
(Z)-3-hexenyl acetate	m
6-methyl-5-hepten-2-one	m
hexanol	m
(Z)-3-hexenol	m
(E,Z)-2,6-dimethyl-1,3,5,7-octatetraene	*)
(E,E)-2,6-dimethyl-1,3,5,7-octatetraene	*)
decanal	m
benzaldehyde	m
linalool	m
methyl benzoate	m
ethyl benzoate	m
estragole	11.1
neral	m
α-terpineol	m
benzyl acetate	m
geranial	3.3
methyl salicylate	m
geranyl acetate	3.8
(Z,E)-4,8,12-trimethyl-1,3,7,11-tridecatetraene	m
(E,E)-4,8-12-trimethyl-1,3,7,11-tridecatetraene	4.3
(E,Z)-2,6-dimethyl-3,5,7-octatrien-2-ol	*)
(E,E)-2,6-dimethyl-3,5,7-octatrien-2-ol	*)
geraniol	12.5
anis aldehyde	m
methyl eugenol	3.1
(E)-nerolidol	12.2
methyl (E)-cinnamate	m
eugenol	2.6
(Z,E)-farnesal	2.0
methyl anthranilate	1.7
(E,E)-farnesal	3.5
chavicol	3.9
(E,E)-farnesol	m
indole	m

Zygopetalum crinitum (cf. p.120)

K 17182, BGSG, April 6, 1991, 12.30–14.30, RK, a.

Compound	Area %
myrcene	1.0
limonene	m
(Z)-ocimene	1.0
(E)-ocimene	28.2
hexyl acetate	m

Appendix

References

1 Hampton, F.A., *The Scent of Flowers and Leaves*. Dulan & Company Ltd., London, 1925.

2 Wilder, L.B. *The Fragrant Garden*. Dover, New York, 1974, p. 149. (Republication of *The Fragrant Path*. Macmillan, 1932.)

3 Plenzat, F., *Duftende Pflanzen in Garten und Haus*. Plenzat, Frankfurt, 1983.

4 Müller A., ‹Natürliche Geruchsskizzen› : Orchideen als Beispiel. *Deutsche Parf.-Zeitung,* 26 (1940), 191, 198, 200.

5 Richardson, D., Fragrance in orchids. *Am. Orchid Soc. Bull.,* 33 (1964), 374–376.

6 Schnepper, J.W., Fugacious fragrance. *Am. Orchid Soc. Bull.,* 35 (1966), 741–743.

7 Schwob, R., The perfume of orchids. *Dragoco Report,* 1979/2, 48–58.

8 Soule, L.C., Fragrance in orchids. *Am. Orchid Soc. Bull.,* 59 (1990), 701–702.

9 Nakamura, S., Tokuda, K., Omata, A., Japan Prize Fragrance Competition. *Am. Orchid Soc. Bull.,* 59 (1990), 1031–1036.

10 Darwin, C., *The Various Contrivances by which Orchids are Fertilized by Insects*. 2nd ed. D. Appleton, New York, 1877 (only seen as summary).

11 Pijl, L. van der, Dodson, C.H., *Orchid Flowers: Their Pollination and Evolution*. Univ. Miami Press, Coral Gables, Fla., 1969, e.g. pp. 83–87 and 128.

12 Williams, N.H., The biology of orchids and euglossine bees. Chapter 4 in *Orchid Biology, Reviews and Perspectives,* II., ed. J. Arditti. Comstock Publishing Associates, Cornell University Press, Ithaca and London, 1982, pp. 120–171.

13 Williams, N.H., Floral fragrances as cues in animal behavior. Chapter 3 in *Handbook of Experimental Pollination Biology*. Scientific and Academic Editions, New York, Cincinnati, Toronto, London, Melbourne, 1983, pp. 50–72.

14 Williams, N.H., Whitten, W.M., Orchid floral fragrances and male euglossine bees: methods and advances in the last sesquidecade. *Biol. Bull.,* 164 (1983), 355–395.

15 Bicchi, C., Joulain, D., Headspace-gas chromatographic analysis of medicinal and aromatic plants and flowers. *Flavour and Fragrance Journal,* 5 (1990), 131–145.

16 a Kaiser, R., Trapping, investigation and reconstitution of flower scents. Chapter 7 in *Perfumes: Art, Science and Technology,* ed. P.M. Müller, D. Lamparsky. Elsevier Applied Science, London, New York, 1991, pp. 213–250.
 b Thiboud, M., Empirical classification of odours. Chapter 8 of the same book, pp. 253–286.

17 Matile, P., Altenburger, R., Rhythms of fragrance emission in flowers. *Planta,* 174 (1988), 242–247.

18 Altenburger, R., Matile, P., Circadian rhythmicity of fragrance emission in flowers of *Hoya carnosa* R. Br. *Planta,* 174 (1988), 248–252.

19 Borg-Karlson, Anna-K., Chemical and ethological studies of pollination in the genus *Ophrys*. *Phytochemistry,* 29 (1990), 1359–1387.

20 Amoore, J.E., Odor theory and odor classification. Chapter 2 in *Fragrance Chemistry – The Science of the Sense of Smell,* ed. E.T. Theimer. Academic Press, New York, London, 1982, pp. 27–76.

21 Ohloff, G., Beschreibung und Klassifizierung von Geruchseindrücken. Chapter 4 in *Riechstoffe und Geruchssinn – Die molekulare Welt der Düfte*. Springer, Berlin, New York, 1990, pp. 71–76.

22 Brunke, E.-J., Schatkowski, D., Struwe, H., Tumbrink, L., Bergamotol and Spirosantalol – New constituents of Eastindian sandalwood oil. In *Flavors and Fragrances: a World Perspective*. Proc. 10th Internat. Congress of Essential Oils, Fragrances and Flavors, Washington, DC. 1986, ed. B.M. Lawrence, B.D. Mookherjee, B.J. Willis. Elsevier Science Publishers, Amsterdam, 1988, pp. 819–831.

References

23 Kaiser, R., Lamparsky, D., Caryophyllane-2,6-β-oxide, a new sesquiterpenoid compound from the oil of *Lippia citriodora* Kunth. *Helv. Chim. Acta,* 59 (1976), 1803–1808.

24 Hawkes, A.D., *Encyclopaedia of Cultivated Orchids.* Faber & Faber, London, 1965.

25 Williams, N.H., Floral fragrance components of *Brassavola. Selbyana,* 5 (1981), 279–285.

26 Byers, J.A., Birgersson, G., Löfquist, J., Bergström, G., Synergistic pheromones and monoterpenes enable aggregation and host recognition by a bark beetle, *Pityogenes chalcographus. Naturwiss.,* 75 (1988), 153–155.

27 Gerlach, G., Schill, R., Composition of orchid scents attracting euglossine bees. *Bot. Acta,* 104 (1991), 379–391.

28 Gerlach, G., Schill, R., Fragrance analyses, an aid to taxonomic relationships of the genus *Coryanthes. Pl. Syst. Evol.,* 168 (1989), 159–165.

29 Silverstein, R.H., Rodin, J.O., Wood, D.L., Sex attractants in frass produced by male *Ips confusus* in ponderosa pine. *Science,* 154 (1966), 509–510.

30 Whitten, W.M., Hills, H.G., Williams, N.H., Occurrence of ipsdienol in floral fragrances. *Phytochemistry,* 29 (1988), 2759–2760.

31 Williams, N.H., Whitten, W.M., Identification of floral fragrance components of *Stanhopea embreei* and attraction of its pollinators to synthetic fragrance compounds. *Am. Orchid. Soc. Bull.,* 51 (1982), 1262–1266.

32 Kaiser, R., New volatile constituents of *Jasminum sambac* (L.) Aiton. In *Flavors and Fragrances: a World Perspective,* Proc. 10th Internat. Congress of Essential Oils, Fragrances and Flavors, Washington, DC. 1986, ed. B.M. Lawrence, B.D. Mookherjee, B.J. Willis. Elsevier Science Publishers, Amsterdam, 1988, pp. 669–684.

33 Lindquist, N., Battiste, M.A., Whitten, W.M., Williams, N.H., Strekowski L., trans-Carvone oxide, a monoterpene epoxide from the fragrance of *Catasetum. Phytochemistry,* 24 (1985), 863–865.

34 Whitten, W.M., Williams, N.H., Armbruster, W.S., Battiste, M.A., Strekowski, L., Lindquist, N., Carvone oxide: An example of convergent evolution in euglossine pollinated plants. *Systematic Botany,* 11 (1986), 222–228.

35 Cerfontain, H., Geenevasen J.A.J., The low temperature photochemistry of (Z)-β-ionone and its photo-isomer. *Tetrahedron,* 37 (1981), 1571–1575.

36 Winter, M., Näf, F., Furrer, A., Pickenhagen, W., Giersch, W., Meister, A., Willhalm, B., Thommen, W., Ohloff, G., (Z)-4,7-Octadiensäureäthylester und (Z)-Buttersäure-3,5-hexadienylester, zwei neue Aromastoffe der roten Passionsfrucht. *Helv. Chim. Acta,* 62 (1979), 135–139.

37 Näf, F., Decorzant, R., Willhalm, B., Velluz, A., Winter, M., Structure and synthesis of two novel ionones identified in the purple passionfruit. *Tetrah. Lett.,* 1977, 1413–1416.

38 Withner, C.L., *The Cattleyas and Their Relatives. Volume II. The Laelias.* Timber Press, Portland, Oregon, 1990, p. 79.

39 Thien, L.B., Bernhardt, P., Gibbs, G.W., Pellmyr, O., Bergström, G., Groth, I., McPherson, G., The pollination of *Zygogynum* (Winteraceae) by a moth, *Sabatinca* (Micropterigidae): An ancient association? *Science,* 227 (1985), 540–543.

40 Francke, W., Heemann, V., Gerken, B., Renwick, J.A.A, Vité, J.P., 2-Ethyl-1,6-dioxaspiro[4,4]-nonane, principal aggregation pheromone of *Pityogenes chalcographus* (L.). *Naturwiss.,* 64 (1977), 590–591.

41 Swoboda, P.A.T., Peers, K.E., Volatile odorous compounds responsible for metallic, fishy taint formed in butterfat by selective oxidation. *J. Sci. Fd. Agric.,* 28 (1977), 1010–1018.

42 Karahadian, C., Lindsay, R.C., Role of oxidative processes in the formation and stability of fish flavors. Chapter 6 in *Flavor Chemistry, Trends and Developments,* ACS Symposium Series 388, ed. R. Teranishi, R.G. Buttery, F. Shahidi, American Chemical Society, Washington, DC., 1989, pp. 60–75.

References

43 Stewart, J., Die Aerangis-Verwandtschaft. *Die Orchidee,* 33 (1982), 48–57.

44 Boland, W., Gäbler, A., Biosynthesis of homoterpenes in higher plants. *Helv. Chim. Acta,* 72 (1989), 247–253.

45 Kaiser, R., 1977, not published.

46 Heinzer, F., Chavanne, M., Meusy, J.P., Maitre, H.P., Giger, E., Baumann, T.W., Ein Beitrag zur Klassifizierung der therapeutisch verwendeten Arten der Gattung Echinacea. *Pharm. Acta Helv.,* 63 (1988), 132–136.

47 Kaiser, R., Olfactory and chemical characteristics of floral scents. Paper presented at the 196th National ACS Meeting, Symposium on Progress in Essential Oil Research, Los Angeles, CA., September 1988.

48 Bergström, G., Groth, I., Pellmyr, O., Endress, P.K., Thien, L.B., Hübener, A., Francke, W., Chemical basis of a highly specific mutualism: Chiral esters attract pollinating beetles in *Eupomatiaceae. Phytochemistry,* 30 (1991), 3221–3225.

49 Gerlach, G., The genus *Coryanthes* in Colombia. *Orquideologia,* 18 (1991), 3–20.

50 Joulain, D., The composition of the headspace from fragrant flowers. *Flavour and Fragrance Journal,* 2 (1987), 149–155.

51 Vogel, S., Duftdrüsen im Dienste der Bestäubung; über Bau und Funktion der Osmophoren. *Abh. Math. Nat. Kl. Ak. Wiss. Mainz,* 10 (1962), 601–763, esp. p. 691.

52 Du Puy, D., Cribb, P., *The Genus Cymbidium.* Christopher Helm, London, 1988; Timber Press, Portland, Oregon, 1988.

53 Schenk, H.P., Lamparsky, D., Analysis of nutmeg oil using chromatographic methods. *J. Chromatogr.,* 204 (1981), 391–395.

54 Flath, R.A., Mon, T.R., Lorenz, G., Whitten, C.J., Mackley, J.W., Volatile components of *Acacia* sp. blossoms. *J. Agric. Food Chem.,* 31 (1983), 1167–1170.

55 Kaiser, R., New volatile constituents of the flower concrete of *Michelia champaca* L. *J. Ess. Oil Res.,* 3 (1991), 129–146.

56 Dalpethado, M.E., *Vanda tessellata. Am. Orchid Soc. Bull.,* 33 (1964), 647–648.

57 Nilsson, L.A., Characteristics and distribution of intermediates between *Platanthera bifolia* and *P. chlorantha* in the Nordic countries. *Nord. J. Bot.,* 5 (1985), 407–419.

58 Tollsten, L., Bergström, G., 1991, private communication. Submitted to *Nord. J. Bot.*

59 Wakayama, S., Namba, S., Hosoi, K., Ohno, M., The synthesis and the absolute configuration of lilac alcohols, new naturally occurring odorous ingredients of lilac flower. *Bull. Chem. Soc. Japan,* 46 (1973), 3183–3187.

60 Wakayama, S., Namba, S., Lilac aldehydes. *Bull. Chem. Soc. Japan,* 47 (1974), 1293–1294.

61 Grob, K., Organic substances in potable water and its precursor. *J. Chromatogr.,* 84 (1973), 255–327.

62 Grob, K., Zürcher, F., Stripping of trace organic substances from water; equipment and procedure. *J. Chromatogr.,* 117 (1976), 285–294.

63 Etzweiler, F., Computer controlled micropreparative isolation and enrichment from GC capillary columns and subsequent sample handling. *HRC & CC,* 11 (1988) 449–456.

64 Maurer, B., Hauser A., New sesquiterpenoids from clary sage oil. *Helv. Chim. Acta,* 66 (1983), 2223–2235.

65 Srinivasan, V., Warnhoff, E.W., Base-catalyzed intramolecular displacements on certain 1,2-epoxides. *Can. J. Chem.,* 54 (1976), 1372–1382 and ref. cited therein.

66 Nikishin, G.J., Spektor, S.S., Glukhovtsev, V.G., *Izv. Akad. Nauk SSSR,* Ser. Khim, 1971, 389 [*Chem. Abstr.* 75 (1971), 151 322 f.].

67 Tanaka, S., Yasuda, A., Yamamoto, H., Nozaki, H., A general method for the synthesis of 1,3-dienes. Simple syntheses of β- and trans-α-farnesene from farnesol. *J. Amer. Chem. Soc.,* 97 (1975), 3252–3254.

References

68 Hue, R., Jubier, A., Andrieux, J., Resplandy, A., Synthèse de phénols polyméthoxylés par la reaction de Baeyer-Villiger. *Bull. Soc. Chim. France,* 1970, 3617–3624.

69 Kitahara, T., Takagi, Y., Matsui, M., Structure and synthesis of novel constituents of yudzu peel oil and their conversion to related monoterpenes. *Agric. Biol. Chem.,* 44 (1980), 897–901.

70 Demole, E., Demole, C., Enggist, P., A chemical investigation of the volatile constituents of Eastindian sandalwood oil. *Helv. Chim. Acta,* 59 (1976), 737–747.

71 European Patent No. 6616 (priority CH 28 June, 1978) assigned to Givaudan.

72 Vogel, S., Pilzmückenblumen als Pilzmimeten. *Flora,* 167 (1978), 329–389.

73 Kaiser, R., Lamparsky, D., Nouveaux constituants de l'absolue de jacinthe et leur comportement olfactif. *Parf. Cosm. Arômes,* 17 (1977), 71–79.

74 a Schmaus, G., 1992, private communication. I should like to thank Dr G. Schmaus, DRAGOCO Gerberding & Co. GmbH, D-Holzminden, for kindly giving me this information in advance.

b Brunke, E.-J., Hammerschmidt, F.J., Schmaus, G., Flower scent of traditional medicinal plants. Paper presented at the 203rd National ACS Meeting, Symposium on Volatile Attractants from Plants, San Francisco, Ca., April 1992. Submitted for publication in the forthcoming symposium book.

Suggestions for further reading

Arditti, J., *Orchid Biology, Reviews and Perspectives*. I and II, Cornell Univ. Press, Ithaca, London, 1977.

Bechtel, H., *Wunderbare und geheimnisvolle Welt der exotischen Orchideen*. Bertelsmann, Gütersloh, 1977.

Bechtel, H., Cribb, P., Launert, E., *Orchideenatlas*. Ulmer, Stuttgart, 1985.

Bockemühl, L., *Odontoglossum*. Brücke-Verlag, Kurt Schmersow, Hildesheim, 1989.

Braem, G.J., *Cattleya*. Part I, *The Brazilian Bifoliate Cattleyas*, Part II, *The Unifoliate Cattleyas*. Brücke-Verlag, Kurt Schmersow, Hildesheim, 1986.

Braem, G.J., *Paphiopedilum*. Brücke-Verlag, Kurt Schmersow, Hildesheim, 1988.

Comber, J.B., *Orchids of Java*. Bentham-Moxon Trust, Royal Botanic Gardens, Kew, 1990.

Cribb, P., *The Genus Paphiopedilum*. A Kew Monograph, Collinridge Books/The Hamlyn Pub. Group, 1987.

Cribb, P., *Flora of Tropical East Africa, Orchidaceae,* Part II. London, 1984.

Cribb, P., *Flora of Tropical East Africa, Orchidaceae,* Part III. London, 1989.

Dressler, R., *The Orchids. Natural History and Classification*. Harvard Univ. Press, Cambridge, Massachusetts, London, 1981.

Du Puy, D., Cribb, P., *The Genus Cymbidium*. Timber Press, Portland, 1988.

Dunsterville, G.C.K., Garay, L., *Venezuelan Orchids Illustrated*. Parts 1–6, 1959–1976.

Dunsterville, G.C.K., Garay, L., *Orchids of Venezuela – an Illustrated Field Guide*. Botanical Museum, Harvard Univ., Cambridge, Mass., 1979.

Hawkes, A.D., *Encyclopaedia of Cultivated Orchids*. Faber & Faber, London, 1965.

Hillerman, F.E., Holst, A.N., *An Introduction to the cultivated Angraecoid Orchids of Madagascar*. Timber Press, Portland, 1986.

Hirmer, M., *Wunderwelt der Orchideen*. Hirmer, Munich 1974.

Holttum, R.E., *Orchids of Malaya,* Ed. 3. In *A Revised Flora of Malaya*. Government Printing Office, Singapore, 1964.

Keller, J., *Intriguing Masdevallias*. HGH Publications, Berkshire, 1984.

Kramer, J., *The World Wildlife Fund Book of Orchids*. Abbeville, New York, 1989.

Luer, C.A., *The Native Orchids of the United States and Canada*. New York Botanical Garden, 1975.

Luer, C.A., *Thesaurus Masdevalliarum*. Helga Königer, Munich, 1983.

Northen, R.T., *Miniature Orchids*. Van Nostrand Reinhold Company, New York, 1980.

Pabst, G.F.J., Dungs., F., *Orchidaceae Brasiliensis*. 2 Parts, Brücke-Verlag, Kurt Schmersow, Hildesheim, 1975–1977.

Pijl, L. van der, Dodson, C.H., *Orchid Flowers: Their Pollination and Evolution*. Univ. Miami Press, Coral Gables, Fla., 1969.

Reinhard, H.R., Gölz, P., Ruedi, P., Wildermuth, H., *Die Orchideen der Schweiz und angrenzender Gebiete*. Fotorotar AG, Egg, Switzerland, 1991.

Sprunger, S., *Orchids from the Botanical Register 1815–1847*. Birkhäuser, Basel, 1991.

Sprunger, S., *Orchideentafeln aus Curtis's Botanical Magazine*. Eugen Ulmer, Stuttgart, 1986.

Suggestions for further reading

Stewart, J., Lindler, H.P., Schelpe, E.A., Hall, A.V., *Wild Orchids of South Africa.* MacMillan South Africa Ltd., 1983.

Summerhayes, V.S., *Flora of Tropical East Africa, Orchidaceae,* Part I. London, 1968.

Withner, C.L., *The Orchids – Scientific Studies.* Wiley-Interscience, New York, 1974.

Withner, C.L., *The Cattleyas and their Relatives.* Vol. I, *The Cattleyas;* Vol. II, *The Laelias,* Timber Press, Portland, 1988.

Die Orchidee, since 1949, Publisher Deutsche Orchideen-Gesellschaft.

American Orchid Society Bulletin, since 1932, Publisher American Orchid Society.

Lindleyana, – the scientific journal of the American Orchid Society, Vol. I, 1986.

Orchid Digest, since 1937. Edition c/o Mrs N.H. Atkinson, Member Secy., Box 916, 95609 Carlmichael CA.

Index of orchid species discussed
(Illustrations are indicated by asterisks*)

Acacallis cyanea Lindl.	49*	*Brassia verucosa* Lindl.	60,61*
Acineta humboldtii	50	*Bulbophyllum barbigerum* Lindl.	147*
Acineta superba (H.B.K.) Rchb.f.	50*	*Bulbophyllum lobbii* Lindl.	146*
Aerangis appendiculata (De Wild.) Schltr.	123,124*,129	*Catasetum pileatum* Rchb.f.	82*,83
Aerangis biloba (Lindl.) Schltr.	124*	*Catasetum viridiflavum* Hk.	83*
Aerangis brachycarpa (A.Rich.) Dur. et Schinz	124,125*,127	*Cattleya araguaiensis* Pabst	66*
		Cattleya bicolor Lindl.	67,68*
Aerangis confusa J. Stewart	125,126*	*Cattleya bicolor* var. *brasiliensis*	68
Aerangis distincta J. Stewart	127,128*	*Cattleya dowiana* Batem. et Rchb.f.	63
Aerangis fastuosa (Rchb.f.) Schltr.	127,128*	*Cattleya granulosa* var. *schofieldiana* (Rchb.f.) Veitch.	68,69*,76
Aerangis friesiorum	125	*Cattleya granulosa* Lindl.	68
Aerangis kirkii (Rolfe) Schltr.	128,129*	*Cattleya guttata* Lindl.	69
Aerangis kotschyana (Rchb.f.) Schltr.	129*	*Cattleya labiata* Lindl.	22*,25,44,62*
		Cattleya lawrenceana Rchb.f.	63*,65
Aerangis somalensis (Schltr.) Schltr.	129,130*	*Cattleya leopoldii* Lem.	69,70*
		Cattleya luteola Lindl.	66*,67,73
Aeranthes grandiflora Lindl.	130*	*Cattleya maxima* Lindl.	64*,65
Aerides crassifolia Par. et Rchb.f.	144*	*Cattleya percivaliana* O'Brien	65*,69,84,119
Aerides fieldingii Jennings	144,145*	*Cattleya porphyroglossa* Linden et Rchb.f.	68,70*
Aerides lawrenceae Rchb.f.	145*	*Cattleya rex* O'Brien	35*
Aerides multiflora Roxb.	145	*Cattleya schilleriana* Rchb.f.	70,71*
Ancistrochilus rothschildianus O'Brien	139,140*	*Caularthron bicornutum* (Hook.) Raf.	70,72*
Angraecopsis amaniensis Summerh.	131*	*Chondrorhyncha chestertonii* Rchb.f.	52*
Angraecopsis breviloba Summerh.	131	*Chondrorhyncha lendyana* Rchb.f.	52,53*,66,87,119
Angraecum aporoides Summerh.	132*	*Cirrhaea dependens* (Lodd.) Rchb.f.	72*
Angraecum bosseri Senghas	132,133*	*Cirrhopetalum fascinor* Rolfe	149*
Angraecum distichum Lindl.	132	*Cirrhopetalum gracillium* (Rolfe) Rolfe	31,148*
Angraecum eburneum ssp. *eburneum* Bory	133*	*Cirrhopetalum ornatissimum* Rchb.f.	150
Angraecum eburneum ssp. *superbum* Bory	133,134*	*Cirrhopetalum robustum* Rolfe	32*,148
Angraecum eichlerianum Kraenzl.	134*	*Cochleanthes aromatica* (Rchb.f.) R.E.Schultes et Garay	53*
Angraecum infundibulare Lindl.	134	*Cochleanthes discolor* (Lindl.) R.E.Schultes et Garay	54*
Angraecum sesquipedale Thou.	23*,25,26,29*,30,132		
Anguloa clowesii Lindl.	50*,51	*Cochleanthes marginata* (Rchb.f.) R.E.Schultes et Garay	54*,55
Anguloa uniflora Ruiz et Pavon	51		
Bollea coelestis (Rchb.f.) Rchb.f.	51*	*Coelogyne corymbosa* Lindl.	150
Brassavola digbyana Lindl.	43,59,60*	*Coelogyne cristata* Lindl.	150
Brassavola fragrans	58	*Coelogyne fimbriata* Lindl.	150
Brassavola glauca Lindl.	43,59,60*	*Coelogyne flaccida* Lindl.	150
Brassavola nodosa (L.) Lindl.	58*	*Coelogyne pandurata* Lindl.	150
Brassavola tuberculata Hook.	58,59*	*Coelogyne zurowetzii* Carr.	150*

Index

Constantia cipoensis C.Porto et Brade	73*
Coryanthes leucocorys Rolfe	74*
Coryanthes mastersiana Lehm.	36*
Coryanthes picturata Rchb.f.	75*
Coryanthes speciosa Rchb.f.	75
Coryanthes vieirae Gerlach	76*,78
Cymbidium ensifolium (L.) Sw.	13,14
Cymbidium faberi Rolfe	151
Cymbidium formosanum	151
Cymbidium goeringii (Rchb.f.) Rchb.f.	13,101,151*
Cymbidium harukanran	151
Cymbidium kanran Mak.	151
Cymbidium sinense Willd.	151
Cymbidium virescens Rchb.f.	14*,151
Dendrobium anosmum Lindl.	152,153*
Dendrobium antennatum var. d'albertsii Lindl.	153,154*
Dendrobium arachnites Rchb.f.	162
Dendrobium beckleri F.Mueller	154,155*
Dendrobium bellatulum Rolfe	156
Dendrobium brymerianum Rchb.f.	154,155*
Dendrobium carniferum Rchb.f.	154,156*
Dendrobium chrysanthum Lindl.	158
Dendrobium chrysotoxum Lindl.	156,157*,158
Dendrobium delacourii Guill.	158*
Dendrobium densiflorum Lindl.	158
Dendrobium draconis Rchb.f.	155
Dendrobium fimbriatum Hook.	158
Dendrobium griffithianum Lindl.	158
Dendrobium harveyanum Rchb.	158
Dendrobium lawesii F.Mueller	39*
Dendrobium lichenastrum (F.Mueller) Kraenzl.	159,160*
Dendrobium linguiforme Sw.	152
Dendrobium margaritaceum Finet.	156
Dendrobium moniliforme (L.) Sw.	14,159,161*
Dendrobium monophyllum F.Mueller	159,161*
Dendrobium pugioniforme Cunn.	154,155*
Dendrobium rigidum R.Br.	159
Dendrobium superbum Rchb.f.	153
Dendrobium trigonopus Rchb.f.	161,162*
Dendrobium unicum Seidenf.	162,163*
Dendrobium virgineum Rchb.f.	164*
Dendrobium williamsonii Day et Rchb.f.	155,156*
Dendrochilum cobbianum Rchb.f.	164*,165*
Dendrochilum glumaceum Lindl.	165
Diaphananthe pellucida (Lindl.) Schltr.	134,135*
Diaphananthe pulchella Summerh.	136*
Dichea rodriguesii Pabst	84*
Disa uniflora Berg.	37,38*
Dracula chestertonii (Rchb.f.) Luer	31,34*,104
Dracula chimaera (Rchb.f.) Luer	103*,104
Dryadella edwallii (Cogn.) Luer	104*
Embreea rodigasiana (Claes ex Cogniaux) Dodson	81,82*
Encyclia adenocarpa (La Llave et Lex.) Schltr.	84,85*
Encyclia baculus (Rchb.f.) Dressler et Pollard	86*
Encyclia citrina (La Llave et Lex.) Dressler	86,87*,159
Encyclia fragrans (Swartz) Lemée	87,88*
Encyclia glumacea (Lindl.) Pabst	88,89*
Epidendrum ciliare L.	27,90*
Epidendrum lacertinum Lindl.	90*,91
Epidendrum nocturnum Jacquin	90*,91
Epidendrum secundum Jacquin	37
Epigeneium lyonii (Ames) Summerh.	165*
Eria hyacinthoides (Bl.) Lindl.	118,166*
Gongora armeniaca Lindl.	77*
Gongora cassidea Rchb.f.	77,78*
Gongora galeata (Lindl.) Rchb.f.	78
Gymnadenia conopea (L.) R.Br.	27*,176,177*
Gymnadenia odoratissima (L.) L.C.Rich.	177
Himantoglossum hircinum (L.) Sprengel	118,179,180*
Houlletia odoratissima Lind. ex Lindl. et Paxt.	91*,92
Huntleya heteroclita C.Schweinf.	56*
Huntleya meleagris Lindl.	55*,56,161
Laelia albida Lindl.	92*
Laelia anceps Lindl.	93*,94,95
Laelia autumnalis Lindl.	93*,94,95
Laelia gouldiana Rchb.f.	94*,95
Laelia harpophylla Rchb.f.	39*
Laelia perinii Batem.	95*,171
Liparis viridiflora (Bl.) Lindl.	118,166,167*
Lycaste aromatica (Hook.) Lindl.	96*
Lycaste cruenta (Lindl.) Lindl.	96,97*
Lycaste locusta Rchb.f.	97,98*
Malaxis monophyllos (L.) Sw.	32*
Masdevallia caesia Roezl	32*,99
Masdevallia elephanticeps Rchb.f.	99*,100
Masdevallia estradae Rchb.f.	101*
Masdevallia glandulosa Königer	44,101,102*
Masdevallia laucheana Kraenzl.	103*
Masdevallia striatella Rchb.f.	99*,100
Masdevallia tridens Rchb.f.	100*,119
Maxillaria nigrescens Lindl.	105*,106
Maxillaria picta Hook.	105*,106

Index

Maxillaria tenuifolia Lindl.	106*
Maxillaria variabilis Lindl.	107*
Miltonia regnellii Rchb.f.	108*
Miltonia schroederiana (Rchb.f.) Veitch	109*
Miltonia spectabilis var. *moreliana* Lindl.	108*,109
Miltoniopsis phalaenopsis (Lind. et Rchb.f.) Garay et Dunsterv.	109,110*, 144,171
Miltoniopsis roezlii (Rchb.f.) Godefroy-Lebenf.	110
Neofinetia falcata (Thunb.) H.H.Hu	14,132,167,168*
Nigritella nigra (L.) Rchb.f.	23*,39*,176
Odontoglossum cirrhosum Lindl.	110,111*
Odontoglossum citrosmum Lindl.	113
Odontoglossum constrictum Lindl.	26,112*,115
Odontoglossum cordatum Lindl.	113
Odontoglossum hallii Lindl.	113
Odontoglossum odoratum Lindl.	113
Odontoglossum pendulum (La Llave et Lex.) Batem.	113,114*
Odontoglossum pulchellum Lindl.	113,114*
Oncidium barbatum Lindl.	115
Oncidium curcutum	115
Oncidium divaricatum Lindl.	115
Oncidium forbesii Hook.	115
Oncidium gardneri Lindl.	115
Oncidium gravesianum Rolfe	115
Oncidium guianense Garay	115
Oncidium longipes Lindl. et Paxt.	115*
Oncidium micropogon Rchb.f.	115
Oncidium ornithorhynchum H.B.K.	117*,118,166
Oncidium pusillum (L.) Rchb.f.	115
Oncidium sarcodes Lindl.	115,116*
Oncidium splendidum Duchartre	116
Oncidium tigrinum La Llave et Lex.	115,116*
Oncidium unguiculatum Lindl.	116
Ophrys insectifera L.	18*,37*
Orchis morio L.	13*
Peristeria elata Hook.	118*
Pescatorea cerina (Lindl.) Rchb.f.	56*,57,119
Pescatorea dayana Rchb.f.	57*
Phalaenopsis amabilis (L.) Blume	168
Phalaenopsis fasciata Rchb.f.	169
Phalaenopsis hieroglyphica (Rchb.f.) Sweet	169
Phalaenopsis schilleriana Rchb.f.	169
Phalaenopsis violacea Witte	169*
Platanthera bifolia (L.) L.C.M.	177,179*
Platanthera chlorantha (Custer) Rchb.f.	177,179*
Plectrelminthus caudatus (Lindl.) Summerh.	137*,138,168
Polystachya campyloglossa Rolfe	139,141*
Polystachya cultriformis (Thou.) Sprengel	141,142*
Polystachya fallax Kraenzl.	141,142*
Polystachya mazumbaiensis Cribb et Podzorski	142*
Rangaeris amaniensis (Kraenzl.) Summerh.	138*
Rhynchostylis coelestis Rchb.f.	170*,171
Rodriguezia decora (Lem.) Rchb.f.	119
Rodriguezia secunda H.B.K.	119
Rodriguezia refracta (Lindl.) Rchb.f.	119*
Stanhopea ecornuta Lem.	79*
Stanhopea jenischiana Kramer et Rchb.f.	79*
Stanhopea oculata (Lodd.) Lindl.	80*
Stanhopea tigrina Batem.	80,81*
Trichocentrum tigrinum Lindl. et Rchb.f.	119,120*
Trichoglottis philippinensis Lindl.	171*
Trixspermum arachnites (Bl.) Rchb.f.	172*
Vanda coerulescens Griff.	173*
Vanda denisoniana (Bens.) Rchb.f.	173*,175
Vanda denisoniana var. *hebraica*	175
Vanda tessellata (Roxb.) G.Don	174*,175
Vanilla planifolia G.Jackson	14
Vanilla pompona Schiede	14*
Zygopetalum crinitum Lodd.	22*,120,121*
Zygopetalum mackaii Hook.	122

Acknowledgements

Countless colleagues, friends and acquaintances, both inside and outside Givaudan-Roure Research, have contributed towards the realization of this book, and I should, at this point, like to offer them my heartfelt thanks.

I am particularly grateful to the following:
Within Givaudan-Roure: Dr Joachim Schmid for supervising the GC/MS measurements of all the orchid scents described in the third section of the book, for the numerous discussions concerning the structural elucidation of new natural products and also for the development and preparation of a database program which allowed us to manage and interpret the huge volume of data, and for the printouts of the chemical structures based on the in-house structural database file; Gunther Zarske for implementing the GC/MS measurements; Jean Märki for the NMR investigations and related discussions; Dr Ernst Billeter for his constructive criticism in a wide variety of technical matters and discussions concerning the structural elucidations; Ernst Scholz for Figure 8; Franz Etzweiler and Erwin Senn for the isolation of new natural products in the µg range using preparative capillary GC; Dr Cornelius Nussbaumer for the synthesis of caryophyll-5-en-2α-ol (23) and the methyl dehydrojasmonates 107 and 144; Dr Daniel Helmlinger for the synthesis of compound 86; Dr Peter Gygax for discussions relating to the synthesis of new natural products; Dr Hanspeter Schenk for assisting with electronic data handling and processing, and for discussions of analytical and synthesis problems; Anita Reolon-Huber for her very valuable and pleasant co-operation during all the initial and developmental stages of this book, ranging from analytical work, the synthesis of new compounds and data processing to the design of formulae diagrams; Ruth Winz for typing up and laying out the manuscript so carefully and efficiently; Dr Dietmar Lamparsky for proofreading and Dr Peter M. Müller, Director of Givaudan-Roure Research, for his constructive criticism of the manuscript, for discussing the basic concept of the project, and for his committed support.

Outside Givaudan-Roure:
Hanspeter Schumacher for his numerous comments, for his standing invitation of many years to carry out investigations in the comprehensive orchid collection at the St. Gallen Botanical Gardens and for Figures 3 and 179; René Volkart, custodian of this collection, for his generous advice and discussions; Waldemar Philipp, custodian of the orchid collection at Zurich University, for discussions and for enabling me to investigate various rare species; Dr Karlheinz Senghas for allowing me to carry out investigations at the orchid collection of Freiburg University (Germany), and Hans Gerhardt Seeger, custodian of this collection for advice and discussions; Dr Günter Gerlach for many discussions and for allowing me to analyse some of the scent samples trapped by him from the orchid collection of Freiburg University and to incorporate the results and the relevant photographs (Figures 38, 71, 72, 91) in this book; Dr Phillip Cribb at Kew Gardens, for his critical appraisal of the nomenclature used and for discussions; Rudolf Jenny for identifying various orchids and for supplying the relevant literature and Figure 42; Professor Bertil Kullenberg, Uppsala University, for providing numerous reprints and for Figure 28; Professor Gunnar Bergström, Göteborg University, for numerous reprints, the exchange of data concerning the *Platanthera* species and for stimulating discussions; Professor Wittko Francke, Hamburg University, for numerous reprints and discussions; Shoji Nakamura, Yokohama, for introducing me to the fascinating world of 'To-Yo-Ran' orchids;

Acknowledgements

Dr Christoph Weymuth for a large number of scent samples which he trapped from his comprehensive private collections and made available for the analytical investigations, and for permission to reproduce the relevant photographs (Figures 45, 63, 64, 65, 69, 70, 79, 80, 81, 82, 98, 115, 126, 127, 136, 138, 187, 189) and for many discussions; Martin Schneider, Head of Editiones Roche, for his very kind co-operation and for the professional production of this book; Sturm, Photolitho AG, Muttenz, for the outstanding reproductions, Morf & Co AG, Basel, for the high-quality printing, and BMP Translation Services, Basel, for the professional English translation. Finally, my heartfelt thanks are due to my wife Lisbeth and my sons Philipp and Thomas, who often accompanied me on my excursions and who at times have had to endure my enthusiasm for orchid scents with great patience and understanding.

Roman Kaiser